AIRCRAFT HYDRAULIC SYSTEMS

AN AVIATION MAINTENANCE PUBLISHERS, INC. TRAINING MANUAL

By Dale Crane

Printed in U.S.A.

For sale by: Aviation Maintenance Publishers
P.O. Box 36   1000 College View Drive
Riverton, Wyoming 82501-0036
Tel: 1 (800) 443-9250

International Standard Book Number 0-89100-058-5

Aviation Maintenance Publishers, Inc.
P.O. Box 36   1000 College View Drive
Riverton, Wyoming 82501-0036

Copyright 1975 by Dale Crane
All Rights Reserved
Printed in the United States of America

SUBSCRIPTION ORDER FORM

S.D.#16

AVIATION MECHANICS JOURNAL

P.O. Box 36
Riverton, WY 82501
Toll Free (800)443-9250

Check Job Title and Business/Industry categories applicable to your aviation affiliation.

JOB TITLE
(Check One Only)

Title			Coding Symbol
☐ Owner	C	☐ Engineer	D
☐ Inspector	I	☐ Instructor	B
☐ Svc. Mgr.	A	☐ Mechanic	E
☐ Supervisor	G	☐ Student	H
☐ Co. Officer	J	☐ Pilot	L
☐ Avionics Tech.	F	☐ Librarian	M
☐ Other	K		

(Please specify)_____

BUSINESS OR INDUSTRY
(Check One Only)

Classifications			Coding Symbol
☐ FBO/Rep. Sta.	1	☐ School	7
☐ Airline/Air Taxi	2	☐ Ag Operator	9
☐ Mfg. or Distri.	3	☐ Self-employed	10
☐ Corp. Avia.	4	☐ Avia. Consultants	11
☐ Military	5	☐ Home Builders	12
☐ Govt. City-St-Fed.	6	☐ Flying Club	13
☐ Other	8		

(Please specify)_____

COMPANY NAME _____ P.O.# _____

TELEPHONE _____

SIGNATURE _____

ADDRESS _____

CITY _____ STATE _____ ZIP _____

U.S. RATES:
Issued Monthly
$16.00/One year
$31.00/Two years

FOREIGN RATES:
(Including Mexico & Canada)
$22.00/One year/Surface mail
$47.00/One year/Airmail
(U.S. FUNDS ON U.S. BANKS ONLY)

Please enter my subscription to the Journal for _____ year(s).
☐ Enclosed please find $_____ to cover my order
☐ Please bill me later
☐ I wish to charge my subscription on my credit card
☐ Master Card # _____ Exp. Date _____
☐ Visa Card # _____ Exp. Date _____

	NO POSTAGE NECESSARY IF MAILED IN THE UNITED STATES

BUSINESS REPLY MAIL

FIRST CLASS Permit #8 Riverton, WY 82501

Postage will be paid by:

INTERNATIONAL

Aviation Mechanics Journal

P.O. Box 36
1000 College View Drive
Riverton, Wyoming 82501

SUBSCRIPTION DEPARTMENT

TABLE OF CONTENTS

I. History of Fluid Power Applications 3

II. Basic Laws of Fluid Power 5

    A. Physical Relationships 5

        1. Definitions and Formulas 5

            a. Area 5

            b. Force 6

            c. Pressure 6

            d. Distance 6

            e. Work 6

            f. Volume 6

            g. Power 7

    B. Physical Laws 8

        1. The Law of Conservation of Energy 8

        2. Static Laws 8

            a. Hydrostatic Paradox 8

            b. Pascal's Law 8

            c. Mechanical Advantage 9

        3. Dynamic Law -- Bernoulli's Principle 10

III. Hydraulic Fluids 11

    A. Types of Fluids 11

        1. Vegetable Base Fluid 11

        2. Mineral Base Fluid 11

        3. Synthetic Fluid 11

    B. Contamination Detection and Protection 12

IV. Evolution of the Aircraft Hydraulic System 13

    A. Sealed Brake System 13

i

      B.   Vented Brake System 13

      C.   Single Acting Actuator System 13

      D.   Power Pump Systems 14

            1.   Manual Pump Control Valve 14

            2.   Automatic Unloading Valve 14

            3.   Open Center System 15

V.   Pneumatic Systems 17

      A.   Backup Pneumatic Systems 17

      B.   Low-Pressure Pneumatic Systems 17

      C.   Full Pneumatic Systems 17

VI. Hydraulic System Components 20

      A.   Reservoirs 20

            1.   Nonpressurized Reservoirs 20

            2.   Pressurized Reservoirs 21

      B.   Hydraulic Pumps 21

            1.   Hand Pumps 21

            2.   Power Pumps 22

                  a.   Constant Displacement Pump 22

                  b.   Variable Displacement Pump 24

      C.   Hydraulic Valves 26

            1.   Flow Control Valves 26

                  a.   Selector Valves 26

                  b.   Sequence Valve 26

                  c.   Priority Valve 27

                  d.   Hydraulic Fuses 28

                  e.   Check Valves 29

            2.   Pressure Control Valves 29

                  a.   Relief Valves 29

                  b.   Pressure Regulators 30

                  c.   Pressure Reducer 31

		D.	Accumulators	31
		E.	Filters	32
		F.	Fluid Lines	33
			1. Rigid Lines	33
			2. Flexible Lines	38
			a. Low-pressure Hose	38
			b. Medium-pressure Hose	38
			c. High-pressure Hose	38
			d. Hose of Teflon	38
		G.	Fluid Line Fittings	39
			1. Pipe Fittings	39
			2. AN Flared Fittings	39
			3. MS Flareless Fittings	41
		H.	Fluid Line Installation	43
			1. Rigid Lines	43
			2. Flexible Lines	44
		I.	High-pressure Seals	46
			1. Chevron Seals	46
			2. O-ring Seals	46
			3. Wipers	50
		J.	Actuators	51
			1. Linear	51
			2. Rotary	53
VII.	Aircraft Landing Gear			55
	A.	Shock Absorbers		55
	B.	Wheel Alignment		57
	C.	Nose Wheel Steering and Shimmy Dampers		57
	D.	Retraction Systems		59
VIII.	Aircraft Brakes			62
	A.	Wheel Units		62
			1. Energizing Brake	62

    2. Nonenergizing Brakes 63
      a. Expander Tube Brakes 63
      b. Single Disc Brakes 64
      c. Multiple Disc Brakes 65
  B. Brake Energizing Systems 66
    1. Independent Brake Master Cylinders 66
    2. Boosted Brakes 68
    3. Power Brakes 68
      a. System Operation 68
      b. Power Brake Control Valves 70
      c. Anti-skid System 70
      d. Deboosters 70
      e. Emergency Brake System 71

IX. Aircraft Wheels, Tires, and Tubes 73
  A. Wheels 73
    1. One-piece Drop Center 73
    2. Removable Flange 73
    3. Two-piece Drop Center 74
  B. Tires 74
    1. Tire Construction 75
      a. Beads 75
      b. Body or Carcass 75
      c. Chafing Strips 75
      d. Undertread 75
      e. Fabric Tread Reinforcement 75
      f. Tread 76
      g. Sidewall 76
      h. Liner 76
    2. Tire Types 76
      a. Type III -- Low-pressure 76
      b. Type VII -- Extra High-pressure 76
      c. New Design 76

		3.	Tire Inspection and Repair	76
			a. Tread Wear	76
			b. Retreading	77
		4.	Demounting Tires	78
		5.	Mounting Tires	78
			a. Tube Type	78
			b. Tubeless Type	78
		6.	Inflating Tires	78
	C.	Tubes and Tube Repair		79
	D.	Tire and Tube Storage		79
X.	Hydraulic System Troubleshooting			80
	A.	Basic Principles of Troubleshooting		80
	B.	Troubleshooting Tips and Procedures		80

Glossary 82

Answers to Study Questions 85

Final Examination 88

Answers to Final Examination 91

v

PREFACE

AIRCRAFT HYDRAULIC SYSTEMS is one of a series of specialized study guides prepared for aviation maintenance personnel, to be used with a corresponding set of 35mm slides and recorded tape casettes.

This series is part of a programmed learning course developed and produced by the Aviation Maintenance Publishers, Inc., (AMPI), to improve and promote the aviation maintenance industry through research, communications, and education. This program is part of that effort to improve the quality of education for aviation technicians throughout the world.

The purpose of each AMPI training series is to provide basic and detailed information on the operation and principles of the various aircraft systems and their components. Texts such as this one are background in nature and provide information in which the A&P can build his understanding of aircraft.

Throughout this text, at appropriate points, is included a series of carefully prepared questions and answers to emphasize key elements of the text, and to encourage the individual to continually test himself for accuracy and retention as he progresses.

For best results, the slide and audio portion should be reviewed first, either in the classroom or by individual study, under the direction of an experienced instructor; then this knowledge should be reinforced with that contained in this book.

If you have any questions or comments regarding this program, or any of the many other programs offered by AMPI, simply contact the Sales Dept., Aviation Maintenance Publishers, Inc., P.O. Box 36, 1000 College View Drive, Riverton, Wyoming 82501-0036, or call 1-800-443-9250.

INTRODUCTION

As the speed traveled by our aircraft has increased, so have the demands placed on its auxiliary systems. At first, wheel brakes replaced tail skids, then speeds increased to the point that retractable landing gear replaced the wheel pants. **At the time this AMPI text is written, small** privately owned aircraft have more complex hydraulic systems than did the military aircraft and airliners of a generation ago.

Hydraulically retracted landing gear with automatic extenders, anti-skid brakes, multiple backup systems, all add to the complexity of the systems the A&P technician must service.

The manufacturers of these complex aircraft along with the manufacturers of their various components all furnish technical information that must be followed in minute detail to get the maximum utility from these machines; but a sound knowledge of the principles behind these details is required to get the optimum service. **This AMPI text, AIRCRAFT HYDRAULIC SYSTEMS**, is written to provide a clear and understandable background in the basics of the science of fluid power transmission as it applies to aircraft hydraulic systems.

This book is arranged to present hydraulic systems from the basic physical laws up through the components to the complete hydraulic and pneumatic systems. Fluid lines and fittings as well as such landing gear components as shock struts, wheels, brakes, and tires are covered. The book concludes with a discussion of logical troubleshooting as it applies to hydraulic systems.

At intervals throughout the text, questions are asked for you to check your progress. Be sure to answer each of these questions before going to the next section, as much of this material builds on what has just been covered. Correct answers are found at the end of the text for you to verify your answer. Some of the words used may be new to you, so you will find them defined in the Glossary. Rather than the same definition found in more technical dictionaries, the definitions are more in line with the language of the A&P. The first time words in the Glossary appear in the text, they are marked with an asterisk(*).

For an over-all review of your comprehension of this material, there is included a twenty question, multiple-choice examination.

Woodward

The waterwheel extracts energy from the weight and velocity of water in motion and is one of the earliest forms of fluid power.

Fig. 1

SECTION I:
HISTORY OF FLUID POWER APPLICATIONS

Fluid power* may be thought of as that system in which work* is accomplished by the movement of a fluid*. This fluid may be either compressible, as a gas (pneumatics*), or incompressible, a liquid (hydraulics*).

Since our earliest recorded history, man has used fluid power; first, to move himself or objects from one place to another by floating down a stream. The tremendous power* in fluids in motion was observed when man saw the havoc wreaked by a wind storm or the devastation from a river on a rampage. This power was first harnessed by the waterwheel as early as the first century BC, and later by the windmill.

The overshoot waterwheel which provided power for our early industrial plants has given way to the high-speed turbine* waterwheels in modern hydroelectric generators, which provide much of the electrical energy we use today.

Waterwheels and windmills are examples of fluid power utilizing an open system. Aircraft fluid power systems use a closed system, in which the moving fluid is confined in such a way that its pressure* may be increased. Thus more work can be done by the same amount of fluid.

Closed hydraulic systems are familiar to us in the form of the hydraulic jack we use to lift our automobiles or the adjustable barber or dental chair. In our aircraft factories, hydropresses exert tons of hydraulic force* to form many of the complex sheet metal parts.

Airplanes would be far less efficient if it were not for fluid power devices: Hydraulic brakes allow the pilot control of the airplane on the ground, without requiring a complex mechanical linkage system. Hydraulic retraction systems pull the heavy landing gear into the wheel wells and decrease wind resistance. Hydraulic-boosted controls make flying high-speed jet aircraft possible. And air under pressure is used to break ice off the wing and tail surfaces.

QUESTION:

1. What is the difference between hydraulics and pneumatics?

SECTION II: BASIC LAWS OF FLUID POWER

Fluid power, as any other branch of physics, must conform to certain definite and well defined laws. These laws have been propounded by great and learned physicists, but it remains for the A&P technicians to apply them to our complex transportation system.

A. PHYSICAL RELATIONSHIPS

1. Definitions and Formulas

a. Area*

This is simply the measurement of the surface of a device. It is expressed in square units, commonly square inches or square centimeters.

[1] Rectangle*

For a rectangle or a square*, the area is the product of the length times the width. In the case of a square, $A = L \times W$ is simply $A = S^2$, in which the length of one of the sides is squared (multiplied by itself). All units must be the same; that is, all in inches or centimeters. The area will then be in square units of the same kind.

[2] Triangle*

A triangle has an area exactly one half that of a rectangle, with the same length and width measurements; so its area is:

$$A = \frac{B \times A}{2}$$

B is the base, the same as the length of a

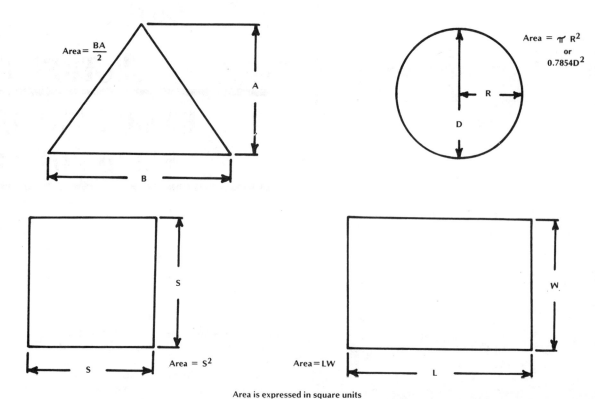

- Fig. 2 -

rectangle, and A is the altitude, the same as the width.

[3] Circle*

The formula for the area of a circle is: $A = \pi R^2$. Pi, π, is a constant*, 3.1416, usually shortened to 3.14; R is the length of the radius (one half of the diameter), which is squared. A more simple formula which does exactly the same thing but which allows us to estimate the area of a circle, is: $A = 0.7854 \times D^2$. Here we have another constant, 0.7854; this is one fourth of pi, (3.1416 ÷ 4). In the first formula we had to square the radius, but in this one we square the diameter, thus saving a step. The square of the diameter is four times the square of the radius, and since we have divided pi by four, we come up with the same results.

0.7854 is close enough to 75%, or 3/4, for purposes of estimation, so to estimate the area of a circle with a ten inch diameter, for example, just find three quarters of the square of the diameter. In this case it is 3/4 of 10^2 (100), or about 75 square inches. This is close enough to the actual area of 78.54 square inches for most estimations.

b. Force

Force is energy exerted, or brought to bear, and is the cause of motion or change. In practical fluid power, it is expressed in pounds, or in metric terms of grams* or kilograms*. Force may be caused by hydraulic or pneumatic action or by a spring.

c. Pressure

When a force is applied over a given area it is termed pressure and is expressed in pounds per square inch or grams per square centimeter. Other units of pressure may be encountered, but, for our purpose, these may all be converted back into basic units of force and area.

d. Distance

In practical mechanics we are concerned with movement, and the distance an object is moved enters into many of our computations. In most power computations it is expressed in feet or meters.

e. Work*

Any time a force causes an object to move, work is done. This work is expressed in foot-pounds or in kilogram-meters.

f. Volume*

Any container for a fluid has a certain volume which may be expressed in cubic units. Volume is usually considered to be the base of the

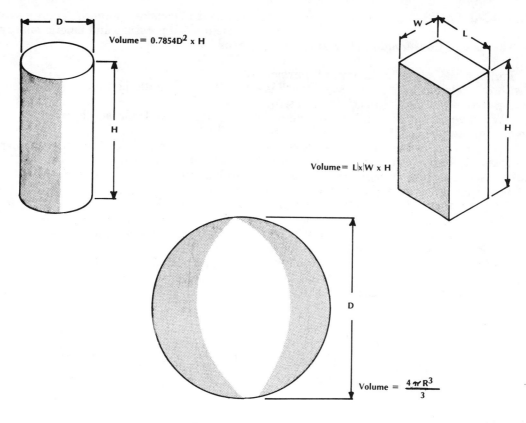

Volume is expressed in cubic units.

- Fig. 3 -

container in square units times its height in the same units. Sometimes fluids are held in spherical containers; whose volume may be found by the formula.

$$V = \frac{4 \pi R^3}{3}$$

in which the volume will be given in cubic units.

g. Power*

Work is simply the product of force and distance, but the amount of power required to accomplish a given amount of work must take into consideration the time required. Power is therefore force times distance, divided by time.

$$\text{Power} = \frac{\text{Force x Distance}}{\text{Time}}$$

A standard unit of power is the horsepower. This is 33,000 foot-pounds of work done in one minute; or, if a shorter period of time is desired, 550 foot-pounds of work done in one second.

In the metric system, a metric horsepower is 4500 kilogram-meters per minute, or 75 kilogram-meters in one second. One metric horsepower is equal to 0.986 horsepower.

In hydraulic systems, power may be computed by considering the flow rate in gallons per minute (one gallon = 231 cubic inches) and the pressure in pounds per square inch, to get force-distance-time relationship. One gallon per minute of flow under a pressure of one pound per square inch will produce 0.000583 horsepower.

Horsepower = Gallons per minute x pounds per square inch x 0.000583

QUESTIONS:

2. What is the relationship between the area of a rectangle and the area of a triangle with the same dimensions?

3. What is the estimated area of a circle having a diameter of five inches?

4. What is the name of the product of force and distance?

5. How many foot-pounds of work per minute is equivalent to one horsepower?

6. How many horsepower is required to pump 30 gallons per minute of hydraulic fluid under a pressure of 1500 psi?

B. PHYSICAL LAWS

1. The Law of Conservation of Energy

Perhaps the most basic law of physics deals with our relationship to energy. Man has not been given the prerogative to create or destroy energy. We have energy at our disposal and are able to change its form, but in the final analysis we end up with exactly the same amount as we started with. In almost any type of mechanical device, we seem to lose energy because of its inefficiency; what actually happens, however, is that we have transformed some of the energy into an unusable form, such as heat from friction.

Basically, we find energy in one of two forms: kinetic* which exists in an object due to its motion; and potential*, which exists in an object because of its position or the arrangement of its parts. In fluid power, a practical way to look at this is to consider potential energy expressed in a fluid as its pressure and kinetic energy expressed as its velocity*.

2. Static Laws

a. Hydrostatic Paradox*

A paradox is a true statement that does not readily appear to be true. In this case, as illustrated in Fig. 4, the static pressure exerted by a column of fluid is proportional to the height of the top of the fluid and is not affected by its volume. What this means is: if we have a container of liquid in any of the shapes of Fig. 4, the pressure indicated by the gage at the bottom of the container will depend on the height of the top of the fluid and will not be affected by the shape of the container or by the amount of fluid.

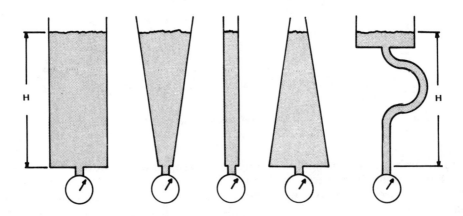

The pressure exerted by a column of liquid is dependent on its height and independent of its volume.

- Figure 4 -

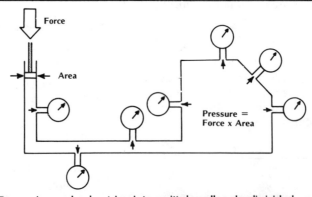

Pressure in an enclosed container is transmitted equally and undiminished to all parts of the container and acts at right angles to the enclosing walls.

- Fig. 5 -

b. Pascal's Law*

Power transmission in a closed hydraulic or pneumatic system is best explained by Pascal's law. Stated in a simple form, it says, "Pressure in an enclosed container is transmitted equally and undiminished to all parts of the container and acts at right angles to the enclosing walls."

In Fig. 5, we see a closed container, filled with

fluid. Pressure gages are arranged around the container to measure the pressure created by force F pushing down on the piston*. The pressure will be F x A pounds per square inch and will be the same on every gage regardless of the position in the system, or the shape of the container. This law is taken advantage of in an automobile brake system. When the brake pedal is depressed, pressure is transmitted equally and undiminished to all of the wheels, regardless of the distance between the wheel and the brake pedal.

obtained with a hydraulic system by applying Pascal's law. In Fig. 7, we have a simple hydraulic jack. The piston on side 1 has an area of one square inch, and on side 2, an area of ten square inches. If a force of ten pounds is applied to side 1, there will be a pressure generated in the system of ten pounds per square inch. According to Pascal's law, this pressure is the same throughout the system and therefore ten pounds of force acts on each square inch of piston 2. This produces a force of 100 pounds which will balance F_2.

Length 1 x Weight 1 = Length 2 x Weight 2

- Fig. 6 -

c. Mechanical Advantage

An application of Pascal's law allows us to see the mechanical advantage we have in a hydraulic system. To briefly review the principle of mechanical advantage, look at the balance in Fig. 6. The product of weight 1 and length 1 makes up what we call moment 1. This force tries to rotate the board counterclockwise and is opposed by moment 2, the product of force 2 and length 2. In a system such as this, it is possible for a small force to lift a large weight. Since we do not get something for nothing, we pay the price in distance moved. The work (force x distance) done by one side of the balance is exactly the same as that done on the other side. For instance, if L_1 is 40 inches and L_2 is 20 inches, W_2 of 100 pounds could be balanced by W_1 of 50 pounds. If W_2 is to be lifted a distance of one foot, W_1 must move two feet. The work done on side 1 is two feet times 50 pounds, or 100 foot-pounds, which is the same as side 2, one foot times 100 pounds, 100 foot-pounds.

The same mechanical advantage may be

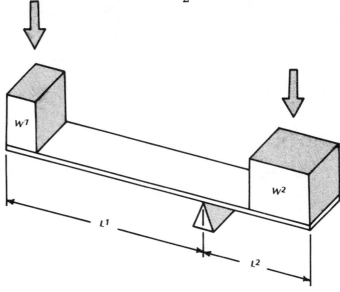

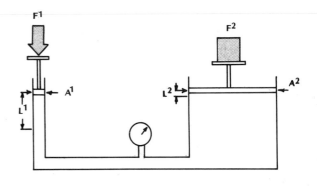

Force 1 x Area 1 = Force 2 x Area 2

- Fig. 7 -

When piston 1 is moved down one inch, one cubic inch of fluid is pushed from side 1 into side 2. This spreads out over the entire area of piston 2 and lifts it only one tenth of an inch. Piston 1 would have to move down ten inches to lift piston 2 one inch.

B. CONTAMINATION DETECTION AND PROTECTION

Hydraulic systems operate with high pressures, and the components have such close fitting parts that any contamination will cause their failure.

When servicing a hydraulic system, be sure that only the proper fluid is used. The service manual of the airplane specifies the fluid, and the reservoir should also be marked with the type of fluid required.

Systems using 5606 fluid may be kept free from contamination by keeping the reservoir tightly closed, and, any time a component is changed, by sealing the lines with the proper fitting caps or plugs. Old 5606 fluid has a somewhat sour smell and is darker in color than fresh fluid. Any fluid suspected of contamination or aging should be drained and the system flushed with fresh fluid. Then, after the flushing fluid is drained, fill the reservoir with fresh fluid. Varsol and Stoddard solvent are compatible with 5606 fluid and may be used to flush the system or to wash components and hoses.

Systems using synthetic fluid are much less tolerant of contamination, and so a special inspection kit including a calibrated microscope may be used to examine this fluid. There are special fittings in the airplane where a sample of fluid may be taken with a hypodermic syringe and diluted with a specific amount of trichlorethylene and squirted into a small filter. The filter is then placed on a microscope slide and examined against a special scale. In this way a qualitative measure may be made of the contamination in the fluid. When there is any doubt with regard to the purity of hydraulic fluid, it should be drained and discarded, the system flushed with trichlorethylene and filled with fresh fluid.

Any time there has been a component failure, contamination is probable, and the manufacturer's recommendation for flushing the system should be followed in detail.

QUESTIONS:

14. How can an A&P technician determine the proper type of hydraulic fluid to use in an airplane?

15. What should be done to a hydraulic system that has encountered a component failure?

SECTION IV: EVOLUTION OF THE AIRCRAFT HYDRAULIC SYSTEM

When flying was less complex, there was very little need for hydraulic systems. The airplanes flew so slow that drag was of no great concern, so the landing gear could hang down in the wind. Landing speeds were so low there was no need for flaps -- once on the ground, the tail skid served as a very effective brake. Paved runways, however, brought out the need for brakes, and the simple hydraulic system came into being.

A. SEALED BRAKE SYSTEM

A simple diaphragm-type master cylinder and expander tube brake, as in Fig. 9, is a complete hydraulic system in its most simple form.

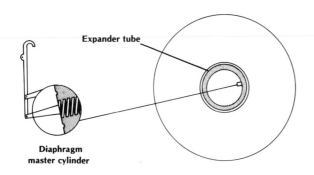

The expander tube brake and diaphragm master cylinder is a hydraulic system of the most simple sort.

- Fig. 9 -

The entire system is sealed, and when the pedal is depressed, fluid is simply moved into the expander tube, which expands and pushes the brake blocks against the drum.

B. VENTED BRAKE SYSTEM

The need to vent the brake system so heat expanding the fluid will not cause the brakes to drag brought about the piston-type master cylinder to replace the sealed diaphragm unit.

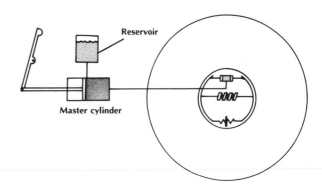

The vented master cylinder provides for expansion of fluid and replenishing lost fluid

- Fig. 10 -

C. SINGLE ACTING ACTUATOR SYSTEM

Landing speeds began to increase, so flaps were added. To lower the flaps hydraulically requires more fluid than one stroke of the pump can supply, so a simple pump and selector valve are added to the single-acting cylinder of Fig. 11. When the selector valve is rotated to the flaps-down position and the hand pump worked, fluid forces the piston out and the flaps down. Air loads on the flap and a spring in the cylinder raise the flaps when the selector valve is rotated to

When the actuation is complete, the pressure will rise and automatically shift the valve into its open center position, allowing fluid to circulate through the system with almost no load on the pump.

QUESTIONS:

16. Are selector valves in a closed center system in series or in parallel with one another?

17. Are the selector valves in an open center in series or in parallel with one another?

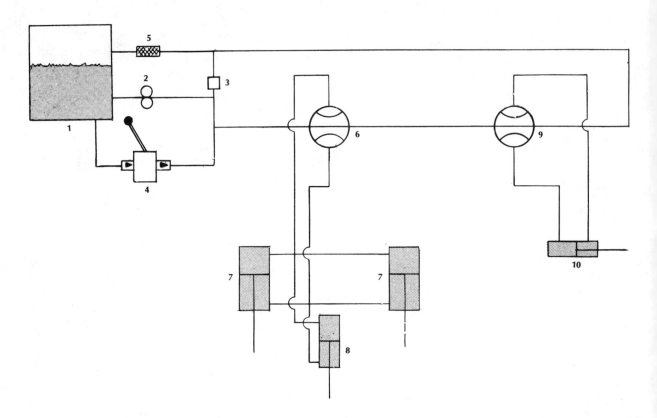

1. Reservoir
2. Engine driven pump
3. Relief valve
4. Hand pump
5. Filter
6. Landing gear selector
7. Main gear actuators
8. Nose gear actuator
9. Flap selector
10. Flap actuator

An open center system requires no separate unloading valve, but returns fluid to the reservoir through the open center of the selector valves when no unit is being actuated.

- Fig. 14 -

SECTION V: PNEUMATIC SYSTEMS

A. BACKUP PNEUMATIC SYSTEMS

In the event of failure of the hydraulic systems in an airplane, there must be provision for emergency extension of the landing gear and the application of the brakes. A very effective and simple system is the pneumatic backup. A steel bottle or cylinder containing approximately 3000 psi of compressed air or nitrogen is installed in the airplane and a shuttle valve in the line to the actuator directs hydraulic fluid for normal actuation, or compressed air for emergency use. If, for instance, there is a failure in the hydraulic system and it is desired to lower the landing gear by the emergency air system, the landing gear selector is put in the gear-down position to provide a return for the hydraulic fluid in the actuators and the emergency air valve is opened. This directs high-pressure air into the actuators and lowers the gear.

B. LOW-PRESSURE PNEUMATIC SYSTEMS

Compressed air under a low pressure is used to drive some of the instruments and inflate the pneumatic de-icer boots. This air pressure is usually provided by an engine-driven vane-type pump. For many years, the main use of these pumps was to drive vacuum-operated instruments, so these pumps are commonly called vacuum pumps. Pneumatic de-icers and air-operated instruments are covered in other AMFI texts and are not discussed in this book, which deals more specifically with systems requiring greater amounts of power.

QUESTION:

18. What two gases may be used for the emergency backup of a hydraulically-retracted landing gear?

C. FULL PNEUMATIC SYSTEMS

The majority of airplanes built in the United States use hydraulic or electric power for heavy duty applications such as landing gear retraction, but compressed air can be used for these systems. Some advantages of compressed air over other systems are:

1. Air is universally available in an inexhaustible supply.

2. The units in a pneumatic system are reasonably simple and lightweight.

3. Compressed air, as a fluid, is lightweight and, since no return system is required, weight is saved.

4. The system is relatively free from temperature problems.

5. There is no fire hazard, and the danger of explosion is minimized by careful design and operation.

6. Installation of proper filters minimizes contamination as a problem.

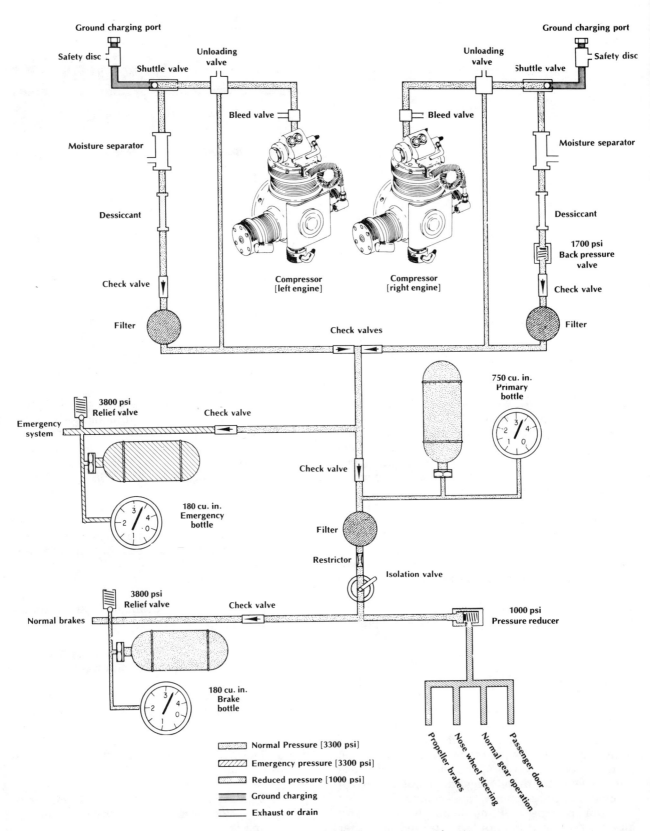

A typical closed center, high-pressure full pneumatic system, such as is found on the Fairchild F-27.

-Fig. 15 -

Fig. 15 is a closed-center, high-pressure pneumatic system, such as is used on the Fairchild F-27 airliner.

The compressors are driven from the accessory gear box of the turbo-prop engines. Air is taken into the first stage through an air duct and compressed, then passed on successively to the other three stages. The discharge air from the fourth stage is routed through an intercooler and a bleed valve to the unloading valve. The bleed valve is kept closed by oil pressure, and, in the event of pressure failure, the valve will open and relieve the pump of any load.

The unloading valve maintains pressure in the system between 2900 and 3300 psi. When the pressure rises to 3300 psi, a check valve traps it and dumps the output of the pump overboard. When the pressure drops to 2900 psi, the output of the pump is directed back into the system. Normal system pressure is in this way maintained at 3300 psi.

A shuttle valve* in the line between the compressor and the main system makes it possible to charge the system from a ground source. When the pressure from the external source is higher than that of the compressor, as it is when the engine is not running, the shuttle slides over and isolates the compressor.

Moisture in a compressed gas system will freeze when the pressure is dropped for actuation, and, for this reason, every bit of water must be removed from the air. A separator collects moisture from the air on a baffle and holds it until the system is shut down. When the inlet pressure to the separator drops below 450 psi, a drain valve opens and all accumulated moisture is discharged overboard. An electric heater prevents water in the separator from freezing.

After the air leaves the water separator with about 98% of its water removed, it passes through a dessicant, or chemical dryer, to remove the last traces of moisture.

The air, before it enters the actual operating system, is filtered through a 10 micron* sintered metal filter. When we realize that the lower level of visibility with the naked eye is about 40 microns, we see that this provides really clean air in the system.

The system in the right engine nacelle has a back pressure valve which is essentially a pressure relief valve in the supply line. This valve does not open until the pressure from the compressor or ground charging system is above 1700 psi, thus assuring that the moisture separator will operate most efficiently. If it is desired to operate the system from an external source below 1700 psi, it can be connected into the left side where there is no back pressure valve.

There are three air storage bottles in this airplane, a 750 cubic inch bottle for the main system, a 180 cubic inch bottle for the normal brake operation, and a second 180 cubic inch bottle for emergency operation of the brakes.

A manually-operated isolation valve allows a technician to close off the air supply so he can service the system without having to discharge the storage bottle.

The majority of the components in this sytem operate with pressure of 1000 psi, so a pressure-reducing valve is installed between the isolation valve and the supply manifold for normal landing gear operation, passenger door, drag brake, propeller brake, and nose wheel steering. This valve not only reduces the pressure to 1000 psi, but also serves as a backup relief valve.

An emergency system stores compressed air under the full system pressure of 3300 psi and supplies it for landing gear emergency extension and emergency brake application.

QUESTIONS:

19. What will happen to the pneumatic system in the event of a failure of the compressor lubrication system?

20. Why is it important that every bit of moisture be removed from the compressed air in a pneumatic system?

21. What is the purpose of the isolation valve?

SECTION VI:
HYDRAULIC SYSTEM COMPONENTS

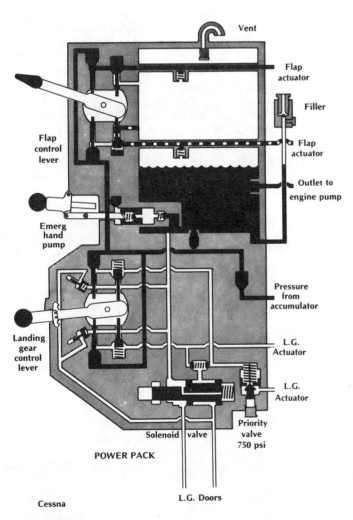

Some of the smaller hydraulic systems have the reservoir, pumps, and valves all in a "power pack" for convenience of servicing.

- Fig. 16 -

A. RESERVOIRS

The hydraulic reservoir not only stores fluid for the system, but serves as an expansion chamber* and a point at which the fluid can purge itself of any air it has accumulated in its operational cycle. Reservoirs must have enough capacity to hold all the fluid that can be returned to the system with any configuration of the gear, flaps, and all other hydraulically-actuated units.

1. Nonpressurized Reservoirs

The fluid return into the reservoir is usually directed in such a way that foaming is minimized and any air in the fluid will be swirled out or extracted. Some reservoirs have filters built into them at the return line so all of the fluid entering the tank is strained.

Reservoirs for most of the medium-size hydraulic systems have two outlets. One, either located partially up the side or connected through a standpipe*. This outlet feeds the engine-driven pump so in the event of a break in a system line or any type of leak that loses all of the pump's fluid, there will still be some fluid in the reservoir. The outlet is near the bottom and the hand pump draws its fluid from here.

There is a trend in the smaller aircraft with limited hydraulic systems to incorporate all of the hydraulic power system into one power pack with the reservoir, valves, and hand pump all in one easy-to-service unit, Fig. 16.

20

2. Pressurized Reservoirs

As airplanes began to fly at higher altitudes where the outside air pressure was low, the returning hydraulic fluid developed a bad tendency toward foaming. To minimize this condition, reservoirs were pressurized. One of the early methods was to inject air into the returning fluid through an aspirator or a venturi tee. The fluid returning into the reservoir flowed through the venturi creating a low pressure, and the air was pulled into its throat. The aerated fluid was swirled into the top of the reservoir where the air was expelled from the liquid. A relief valve on the reservoir maintained a pressure of about 12 psi on the fluid. Some turbine engine-powered aircraft use a small amount of filtered compressor bleed air to pressurize the reservoir.

Jet aircraft that fly at very high altitudes and systems which place heavy demands on the fluid require reservoirs pressurized to a higher pressure. Typical of these is the one used in the Douglas DC-9, Fig. 17.

between the side acted on by the system pressure and that which acts on the fluid in the reservoir. The 3000 psi system pressure, therefore, applies 30 psi pressure on the fluid in the reservoir; the diaphragm assembly moves up, increasing the volume of the reservoir.

When hydraulic fluid is used to perform work, heat is generated. This heat must be dissipated from the fluid, and, in the larger systems, such as the one used in the Boeing 727, heat exchangers are installed in the fuel tanks, Fig. 18.

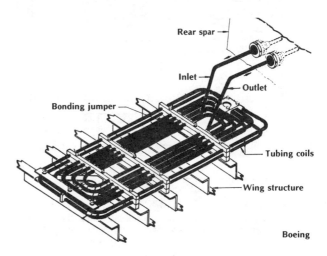

The heat generated in the hydraulic fluid may be removed by a heat exchanger in the bottom of one of the fuel tanks.

- Fig. 18 -

The hydraulic fluid, on the way back into the reservoir, passes through these coils and its heat is given up to the fuel in the tank. Restrictions are made regarding the operation of any of these hydraulic systems on the ground when there is less than a certain amount of fuel in the tank in which these heat exchangers are located.

B. HYDRAULIC PUMPS

Fluid power is available in an aircraft hydraulic system when fluid is moved under pressure. Pumps used in these systems are simply fluid movers, rather than pressure generators. Pressure can be generated only when there is a restriction to the flow of the fluid being moved.

There are two basic types of hydraulic pumps: those operated by hand, and those driven by some source of power, such as by an electric motor or an aircraft engine.

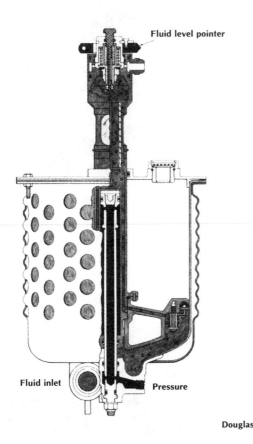

Heavy-duty hydraulic systems require pressurized reservoirs.

- Fig. 17 -

The fluid is pressurized to approximately 30 psi by the system pressure acting on the piston in the reservoir. There is a 1:100 relationship in the area

1. Hand Pumps

Single-action pumps move fluid only on one stroke of the piston, while double-action pumps move fluid with both strokes. Double action pumps are the only ones commonly used in aircraft hydraulic systems because of their greater efficiency. Fig. 19 is a diagram of a piston rod displacement hand pump.

2. Power Pumps

These may be classified as either constant or variable displacement pumps. A constant displacement pump* is one that moves a given amount of fluid each time it rotates. A pump of this type must have some sort of unloading device or regulator to prevent its building up so much pressure that it will rupture a line or perhaps damage the pump itself.

Variable displacement pumps* move a volume of fluid proportional to the demands of the system. These pumps are quite often of the piston type, and their output volume is varied by changing the stroke of the pump.

a. Constant Displacement Pump

One of the simpler constant displacement pumps used to move a rather large volume of fluid under a relatively low pressure is the vane pump, Fig. 20.

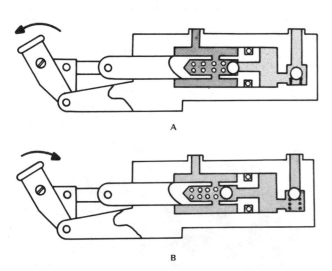

The piston rod displacement hand pump is a double-acting pump. Fluid is pushed out of the pump on every stroke, but taken in only on every other stroke.

- Fig. 19 -

On the stroke in which the piston is pulled out of the cylinder, fluid is drawn in through the inlet check valve and the fluid on the back side of the piston is forced out the pump outlet. When the piston is forced into the cylinder, the rod displaces part of the fluid and some of it is again forced out the discharge.

Let's assume some values: the large end of the piston has an area of two square inches, the rod displaces one square inch, and the piston moves one inch. When the piston moves out of the cylinder, two cubic inches of fluid is drawn in. Now when the piston is moved into the cylinder, the two cubic inches of fluid is forced out, but the space behind the piston has only one cubic inch of volume; so one cubic inch of fluid must be forced out the pump discharge port. When the piston is again pulled out of the cylinder, the remaining one cubic inch is forced out of the pump. Every time the piston is moved out of the cylinder, two cubic inches of fluid is taken in and one cubic inch is discharged. Each time the piston moves into the cylinder, one cubic inch is discharged but no fluid is taken into the pump.

If a force of 500 pounds is exerted on the piston as it is pulled out of the cylinder, a pressure of 500 psi will be built up. On the return stroke, however, since there is a working piston area of two square inches, the same 500 pounds of force will generate only 250 psi pressure.

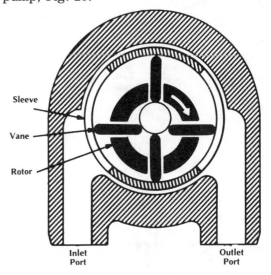

The vane pump is constant displacement and moves a relatively large volume of fluid under relatively low pressures.

- Fig. 20 -

The vanes are free-floating in the rotor and are held against the wall of the steel sleeve by a spacer. As the rotor turns in the direction indicated by the arrow, the volume between the vanes on the inlet side increases and the volume between the vanes on the discharge side decreases. This draws fluid into the pump and forces it out the other side. This type of pump finds wide use in the aircraft for moving fuel for piston engines, and also air for gyro instruments and pneumatic de-icer boots. It is used also to a somewhat lesser degree for hydraulic systems.

To move the medium volumes of fluid under medium pressures required by some hydraulic systems, gear pumps are used. The two types commonly used for aircraft hydraulic systems are the spur gear and the gerotor.

The simple spur gear pump, Fig. 21, uses two meshing gears closely fitted into a housing.

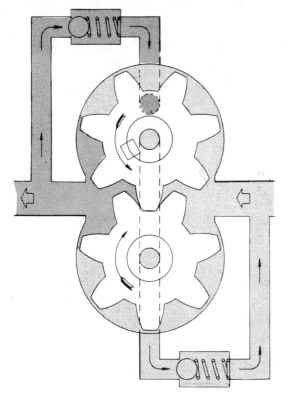

A spur gear pump moves a medium volume of fluid under pressures up to about 1500 psi.

- Fig. 21 -

One of the gears is driven by the engine accessory drive, and this gear drives the other. As the gears rotate in the direction shown by the arrows, the space between the teeth on the inlet side becomes larger. Fluid is pulled into this space, trapped between the teeth and the housing and carried around to the discharge side of the pump. Here, the teeth of the two gears come into mesh, decreasing the volume, and force the fluid out the pump discharge. A small amount of fluid is allowed to leak past the gears and around the shaft for lubrication, cooling, and sealing. This fluid drains into the hollow shafts of the gears and is picked up by the low pressure at the inlet side of the pump. A weak relief valve holds the oil in the hollow shafts until it builds up a pressure of about 15 psi. This so-called case pressure* is maintained so that in the event the shaft or seal becomes scored, fluid will be forced out rather than air being drawn into the pump. Air would otherwise displace the fluid needed for lubrication and the pump would be damaged.

As the output pressure from the gear pump builds up, there is a tendency for the case to distort and allow increased leakage. To prevent this, some pumps have high-pressure oil from the discharge side of the pump fed through a check valve into a cavity behind the bushing flanges. The bushings are thus forced tight against the side of the gears, decreasing the side clearance and minimizing leakage, also compensating for bushing wear.

The gerotor pump* is sort of a combination internal-external gear pump. Fig. 22 shows its operation. Its four tooth spur gear is driven by an accessory drive from the engine, and as it turns it rotates the five tooth internal gear rotor. If you look at the relationship between the two gears, you will see that as the spur gear rotates and turns the internal gear, the space between the teeth gets larger on one side, smaller on the other. Covering these gears is a plate with a crescent shaped opening above each side of the gears. The opening above the space which is getting larger is the inlet side of the pump, and the opening above

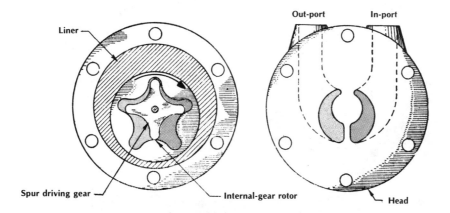

The gerotor pump is a special form of gear pump, producing up to about 1500 psi pressure with a moderate flow.

- Fig. 22 -

the side having the gears coming into mesh is the outlet.

High-pressure, low-volume pumping is often done in aircraft with multiple piston pumps. Fig. 23 illustrates one type of fixed-angle piston pump.

There are usually seven or nine axially-drilled holes in the rotating cylinder block, and fitted into each of these holes are close fitting pistons attached by a ball-jointed rod to a drive plate. The cylinder block and pistons are rotated as a unit by the engine. The housing is angled so the pistons on one side of the cylinder block are at the bottom of their stroke while those on the other side are at the top of theirs. As the pump rotates one half turn, half of the pistons move from the top of their stroke to the bottom; the pistons on the other side move from the bottom of their stroke to the top. A valve plate with two crescent shaped openings covers the end of the cylinders, one opening above the pistons moving up and the other above the piston moving down. As the pistons move down, they pull fluid into the pump, and as they move up, they force this fluid out into the system.

b. Variable Displacement Pump

An unloading valve of some sort is required for a constant displacement pump, but the same force used to control this valve may be used to control the output of a variable displacement pump; so there is no need for a separate control. One of the more popular variable displacement pumps used for high-pressure aircraft systems is the Stratopower demand type pump, Fig. 24. This pump uses nine axially-oriented pistons and cylinders. The pistons are driven up and down in the cylinders by a wedge shaped drive cam, and pistons bear against the cam with ball joint slippers. When the

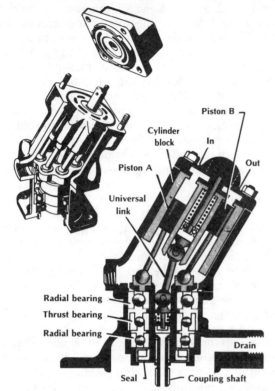

Fixed angle piston pump, a constant displacement pump for high pressures with relatively low flow rates.

- Fig. 23 -

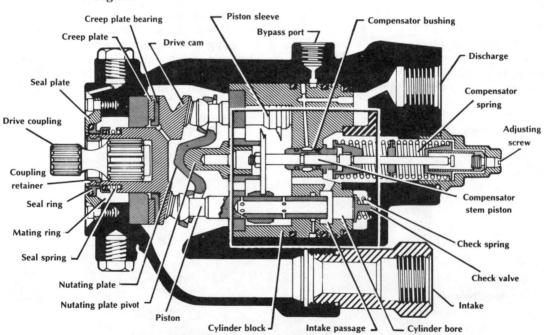

The stratopower demand pump is a variable displacement pump whose compensator varies the EFFECTIVE stroke of the pump.

- Fig. 24 -

thick part of the cam is against the piston, it is at the top of its stroke; and as the cam rotates, the piston moves down the cylinder until, at the thin part of the cam, it is at the bottom. The stroke is the same regardless of the amount of fluid demanded by the system, but the **effective** length of the stroke controls the amount of fluid pumped.

The balance of forces that controls the pressure the pump holds on the system is between the compensator spring and the compensator stem piston. You will notice in Fig. 24 that a passage from the discharge side of the pump directs output fluid pressure around the compensator stem. This stem is cut with a shoulder which serves as a piston. As the system pressure rises, this stem is pushed up, compressing the compensator spring. Attached to the stem is a spider, Fig. 25, which moves the sleeves up or down the pistons.

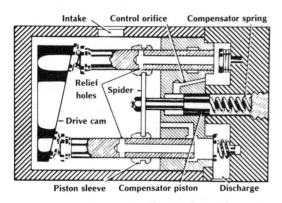

The spider moves the piston sleeves to vary the effective stroke as a function of the balance of forces between the compensator spring and the force on the compensator piston.

- **Figure 25** -

When the pressure is high, Fig. 26-A, it acts on the stem piston to raise the spider against the compensator spring, and the relief holes near the bottom of the pistons are uncovered during all of the stroke. The pistons now stroke up and down, but no fluid is forced out of the pump, as it is all relieved back into the pump. Near the top of the stroke a bypass hole in the piston aligns with a passage in the pump housing and a small amount of fluid is bypassed back into the reservoir, just enough for lubricating and cooling the pump. When the pressure is low, Fig. 26-B, the compensator spring forces the spider and sleeves down the piston, covering the relief hole when the piston is near the bottom of its stroke. In this way, the full stroke of the piston is utilized in moving fluid. Fluid is forced out through the check valves into the pump discharge line. In any condition of intermediate pressure, the sleeve closes the relief holes at some point along the stroke of the piston. In this way enough fluid is pumped to maintain the system pressure at that level for which the compensator spring is set.

QUESTIONS:

22. Name three purposes of a hydraulic reservoir.
23. How is foaming in the hydraulic reservoir of a jet aircraft prevented?
24. What is the purpose of a hydraulic pump? Does it move fluid or generate pressure?
25. In a spur gear type hydraulic pump, does the fluid flow between the gears or around their outside?
26. Why is case pressure held in a spur gear type hydraulic pump?
27. Which type pump is normally used for higher pressure outputs; vane type, gear type, or piston type?
28. Why does a variable displacement hydraulic pump not require a separate unloading valve?
29. What two forces balance in the Stratopower demand-type hydraulic pump to control the pump's output?

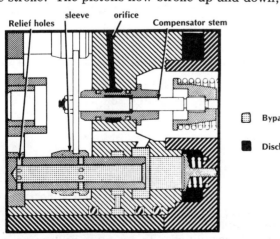
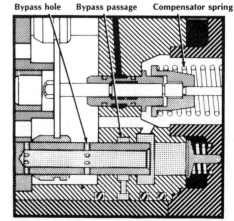

A - The pressure is high; the compensator piston has moved the spider to the top so there is no flow out the discharge.   B - When the pressure is low, the compensator spring moves the spider down and closes the relief hole as soon as the piston starts up. This provides maximum flow.

- **Fig. 26** -

C. HYDRAULIC VALVES

1. Flow Control Valves

a. Selector Valves

One of the more common flow control valves is the selector valve, which determines the direction of flow of fluid to retract or extend the landing gear or to select the position of the flaps. There are two common types of selector valves: the open center valve, which directs fluid through the center of the valve back to the reservoir when a unit is not being actuated, and the closed center valve, which stops the flow of fluid when it is in the neutral position. Both valves direct fluid under pressure to one side of the actuator and vent the opposite side to the reservoir.

For systems using relatively low pressure for actuation, a simple plug-type selector valve, Fig. 27, is often used.

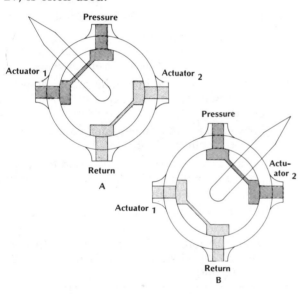

A plug-type rotary selector valve used in closed center hydraulic systems.

- Fig. 27 -

In one position, the pressure port and actuator port 1 are connected. Actuator port 2 is connected to the return line. When the selector handle is turned ninety degrees, the actuator ports are reversed to the pressure and return lines.

Higher pressure systems require a more positive shut-off of fluid flow and a poppet-type selector is often used. In Fig. 28-A the upper right and lower left poppets are off their seat, and the fluid is flowing such a way that the pressure is supplied to actuator lines 2 and line 1 is connected to the return line. When the handle is in the neutral position, all of the poppets are closed, and no fluid flows. When the handle is rotated the opposite position, Fig. 28-B, the pressure and return lines feed the opposite sides of the actuator.

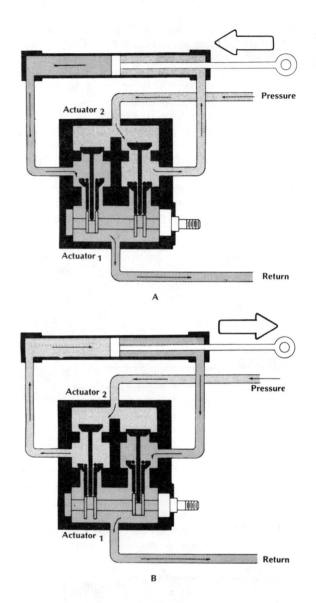

Poppet selector valves may be used in high-pressure closed center systems.

- Fig. 28 -

b. Sequence Valve

Modern aircraft with retractable landing gear often have doors that close in flight to cover the wheel well and make the airplane more streamlined. To be sure the landing gear does not extend while the doors are closed, sequence valves are used. These are actually check valves which allow a flow in one direction but may be opened manually so fluid can flow freely in both directions, Fig. 29.

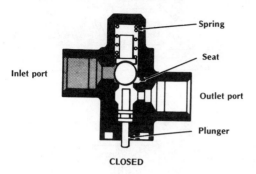

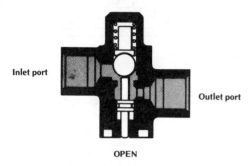

A sequence valve acts as a check valve until its plunger is mechanically pushed up, opening the return path.

- Fig. 29 -

Fig. 30 shows the position in a landing gear actuation system in which these valves would be installed.

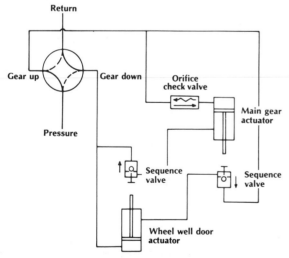

Position of sequence valves in a landing gear retraction system.

- Fig. 30 -

The wheel well doors must be fully open before the sequence valve will allow fluid to flow into the main landing gear cylinder. The return fluid flows unrestricted through the sequence valve on its way back into the reservoir.

c. **Priority Valve**

These valves, Fig. 31, are similar to sequence valves except that they are opened by hydraulic pressure rather than by mechanical contact.

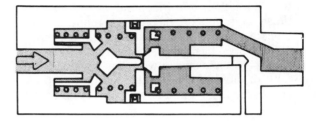

A - Insufficient pressure

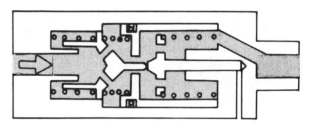

B - Full pressure

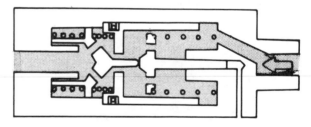

C - Return flow

Priority valves act as check valves until sufficient pressure is built up in the system to open them for full flow.

- Fig. 31 -

They are called priority valves because such devices as wheel well doors, which must operate first, require a lower pressure than the main landing gear, and the valve will shut off all the flow to the main gear until the doors have actuated and the pressure builds up at the end of the actuator stroke. When this build-up occurs, the priority valve opens and fluid can flow to the main gear.

d. Hydraulic Fuses

Modern jet aircraft are dependent on their hydraulic systems, not only for raising and lowering the landing gear, but for control system boosts, thrust reversers, flaps, brakes, and many auxiliary systems. For this reason most aircraft use more than one independent system; and in these systems, provisions are made to fuse or block a line if a serious leak should occur.

Of the two basic types of hydraulic fuses in use, one operates in such a way that it will shut off the flow of fluid if sufficient pressure drop occurs across the fuse. Fluid flows from A to B in Fig. 32.

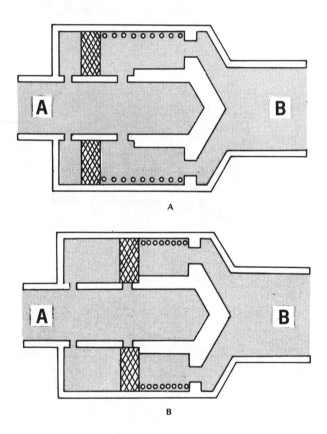

A - Flow rate is low enough that the piston is not moved over against its spring.
B - Flow rate is high enough that the pressure drop across the orifice moves the piston over and stops the flow.

- Fig. 32 -

The spring keeps the passage open and fluid flows into the system for normal operation. If a break should occur in the line, the pressure at side B will drop, and fluid pressure will force the slide over to close the valve and prevent further loss of fluid, Fig. 32-B. Reverse flow is in no way hindered or controlled by this type of fuse.

A second type of fuse, Fig. 33, does not operate on the principle of pressure drop, but it will shut off the flow after a given amount of fluid has passed through the line.

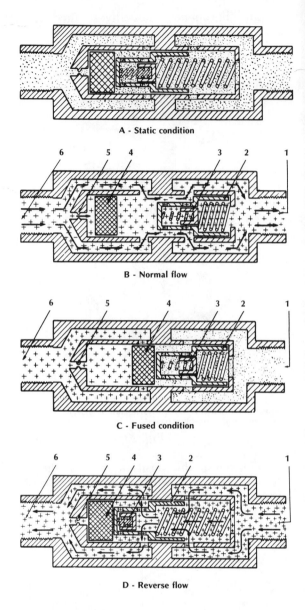

1 - Outlet to actuator unit;   2 - Sleeve valve;   3 - Check valve;   4 - Piston;
5 - Metering orifice;   6 - Inlet from selector valve

- Fig. 33 -

In the static condition, Fig. 33-A, all of the passages are closed off. When fluid begins to flow in the direction of normal operation, Fig. 33-B, the sleeve moves over, compressing the large spring and opening the valve for normal flow. At this time some fluid enters the small orifice and begins to drift the piston over. Normal operation of the unit protected by this fuse does not require enough flow to allow the piston to drift completely over and seal off the line. If there is a leak, however, sufficient fluid will flow that the piston will move over and block the line, as seen in Fig. 33-C. Reverse flow is provided for as the fluid acts on the small piston, compressing its spring and opening the passages for the return fluid, Fig. 33-D.

e. Check Valves

There are many instances in an aircraft hydraulic system where it is desired to allow flow in one direction and prevent it in the opposite direction. This is accomplished by the use of check valves. There are several types of these valves; the ball, cone, and swing valves, Fig. 34 A, B, and C, are the most common.

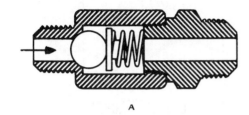

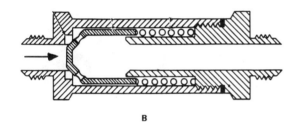

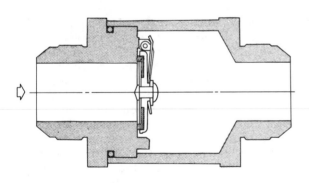

A - Ball check valve; B - Cone check valve; C - Swing check valve

-Fig. 34 -

Certain applications require full flow one way and restricted flow in the opposite direction. An example of this is in a landing gear system where the weight of the gear and air loads causes the extension to be excessively fast, and the weight of the gear against the air loads requires every bit of pressure possible to get the gear up. An orifice check valve* is installed in such a way that fluid flowing into the gear-up lines finds no restriction, and fluid leaving the gear-up side is restricted by the orifice in the check valve, Fig. 35.

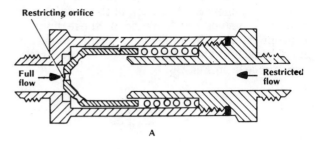

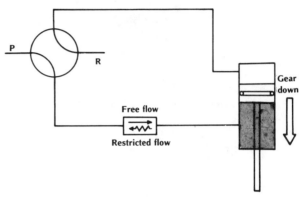

A - Orifice check valve;
B - Orifice check valve, installed in a landing gear system

- Fig. 35 -

QUESTIONS:

30. What is the basic difference between a closed center and an open center selector valve?

31. What is the difference between a sequence valve and a priority valve?

32. On what two principles do hydraulic fuses operate?

33. Would an orifice check valve normally be placed in the up or down line of an actuating cylinder for a landing gear that can free-fall to extend?

2. Pressure Control Valves

a. Relief Valves

The most simple type of pressure control valve is the relief valve. It is used primarily as a backup rather than a pressure control device because of the heat generated and power dissipated when it relieves pressure. System pressure relief valves are set to relieve at a pressure above that maintained by the system pressure regulator, and only in the event of a malfunction of the regulator will the relief valve be called into service. In systems where fluid may be trapped in a line between the actuator and its selector valve there is the problem of pressure build-up by thermal expansion of the fluid. Thermal relief valves are installed in these lines to prevent damage by relieving a small amount of fluid back into the return line. A simple relief valve is seen in Fig. 36.

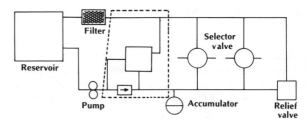

A - Balanced type unloading valve in system

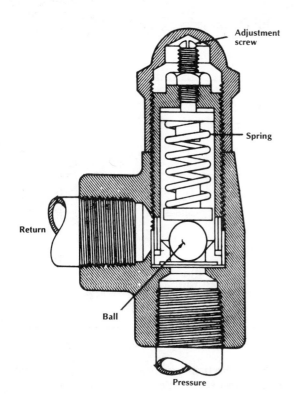

Simple relief valve

- Fig. 36 -

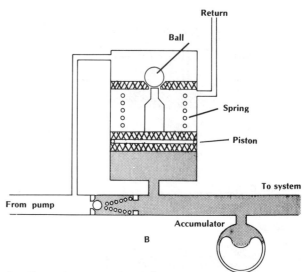

B - Piston area = 1 sq. in.; Ball seat area = 1/3 sq. in.; Spring force = 1000 pounds; Kick-in pressure = 1000 psi; Kick-out pressure = 1500 psi.

- Fig. 37 -

b. Pressure Regulators

Closed-center hydraulic systems require a regulator to maintain the pressure within a specified range and to keep the pump unloaded when no unit is being actuated. The most simple pressure regulator is the balanced type whose principle of operation is shown in Fig. 37.

Starting with a discharged, or flat, system, the pump pushes fluid through the check valve into the system and the accumulator. When no fluid is required for actuation, the accumulator fills and pressure builds up. This pressure pushed up on the piston and also down on the ball. A condition is reached where there is a balance of forces: both the fluid pressure on the ball and the spring act downward and the hydraulic pressure on the piston pushes up. At the condition of balance, when the pressure is 1500 psi, there will be a force of 1500 pounds pushing up on the piston. There will be a total downward force of 1000 pounds applied by the spring, and 1/3 of 1500, or 500 pounds of hydraulic force pushing down on the ball. Now if the pressure rises above this value, since the spring force is constant and not affected by the hydraulic pressure, the piston will move up and unseat the ball. When the ball unseats, the fluid returns to the reservoir and the pump pressure drops to essentially zero. The check valve seats and holds pressure trapped in the accumulator and the system. This unloaded condition will continue until the pressure on the system drops to 1000 psi, at which point the spring can force the piston down and allow the ball to seat. When the ball seats, the pressure again rises to the unloaded pressure.

QUESTIONS:

34. Why is it impractical to use a relief valve to maintain system pressure in a closed center hydraulic system?

35. In a balanced type regulator such as Fig. 37, if the ball seat area is 1/4 sq. in., the piston area is 1 square inch and the spring force is 1500 pounds, what is the kick-in and kick-out pressure?

c. Pressure Reducer

If it is desired to reduce the pressure in some branches of a hydraulic system, a simple pressure reducing valve may be used. The valve shown in Fig. 38 reduces the pressure by the action of a balance of hydraulic and spring forces.

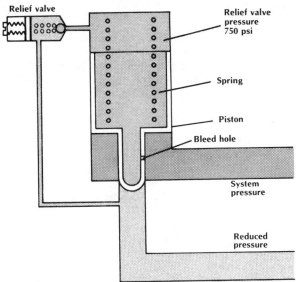

System pressure 1500 psi; Relief valve pressure 750 psi; Spring force 100 pounds; Shoulder area 1/2 sq. in.; Ball seat area = 1/2 sq. in.; Reduced pressure = 200 psi.

Downward force: Spring 100 lb + Hydraulics 750 lb total 850 pounds.
Upward force: Shoulder [1500 ÷ 2] = 750 lb + Ball [200 ÷ 2] = 100 lb, total 850 lb.

Pressure reducing valve
- Fig. 38 -

Let's assume that a piston having an area of one square inch is held against its seat by a spring with 100 pounds of force. The piston has a shoulder area of 1/2 square inch which is acted on by the full 1500 psi system pressure, and a seat area of 1/2 square inch acted on by the 200 psi reduced pressure. A tiny hole in the piston bleeds fluid into the chamber behind the piston and the relief valve maintains this pressure at 750 psi. This relief action is determined by the pressure inside the piston cavity acting on one side of the relief ball and the spring and reduced pressure (200 psi) acting on the opposite side. When the reduced pressure drops, the hydraulic force on the ball drops, allowing it to unseat. This decreases the hydraulic force on the piston and allows it to move up. Fluid now flows into the reduced pressure line and restores the 200 psi. This increased pressure closes the relief valve so that the pressure behind the piston can again come up to 750 psi and seat the valve. Rather than the piston chattering, the tiny bleed hole causes it to have a relatively smooth action, and it remains off its seat just enough to maintain the reduced pressure as it is used.

QUESTION:

36. What is the basic difference between a regulator, a relief valve, and a pressure reducer?

D. ACCUMULATORS

Hydraulic fluid is noncompressible, and in order to store pressure, we must have a compressible fluid. These conflicts are resolved by the use of an accumulator. There are three basic types of accumulators; two are hollow steel spheres, divided into two compartments by either a diaphragm or a bladder; the other is a steel cylinder with a floating piston forming the two compartments. Fig. 39 shows each type.

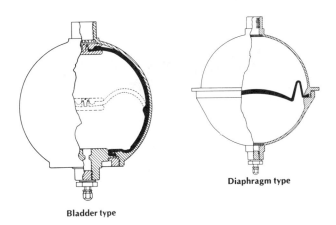

Bladder type   Diaphragm type

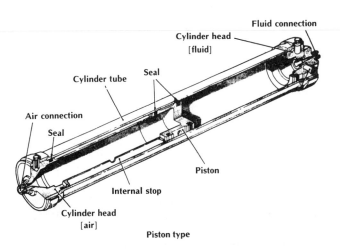

Piston type

ACCUMULATORS

- Fig. 39 -

The accumulator is charged with compressed air or nitrogen to a pressure of approximately one third the system pressure. As the pump forces hydraulic fluid into the accumulator, the air is further compressed and exerts a force on the hydraulic fluid, holding it under pressure after the system pressure regulator has unloaded the pump.

Air valves used in accumulators may be one of three types. The most simple, the AN 812, Fig. 40-A, seals the air in the accumulator with a high-pressure valve core similar in appearance to that used in tires.

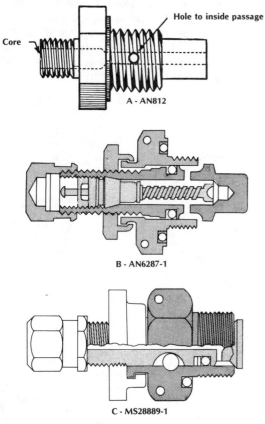

HIGH-PRESSURE AIR VALVES

- Fig. 40 -

These are actually quite different, however, in construction, and high-pressure cores may be identified by the letter "H" embossed on the end of the stem. To deflate an accumulator using an AN 812 valve, loosen the valve body in the accumulator. The hole in the side allows the air to leak past the loosened threads. To deflate an accumulator using an AN-6287-1 valve, loosen the swivel nut one turn and depress the valve stem. For charging, attach the hose to the valve, loosen the swivel nut one turn, and apply the air pressure. Tightening the swivel nut provides a metal-to-metal seal so the hose may be removed and the valve cap replaced.

The MS28889 valve is similar in appearance to the AN 6287 except that there is no valve core and the swivel nut is the same size as the body of the valve (3/4"). One other feature of this valve is the roll pin that prevents the housing being screwed too far into the body.

QUESTIONS:

37. How can hydraulic pressure be stored?

38. How can you tell the difference between the valve core required in an accumulator valve and that used for a tire?

E. FILTERS

One of the more important requirements for hydraulic fluid is its cleanness. Solid particle contaminants must all be removed as they can damage components.

Filtering capability is measured in microns, one micron being one millionth of a meter, or 39 millionths of an inch.

RELATIVE SIZES	
Grain of salt	70 microns
Lower limit of visibility (naked eye)	40 microns
White blood cells	25 microns

SCREEN SIZES		
U.S. Sieve no.	U.S. linear inch	Opening in microns
50	52.36	297
100	101.01	149
200	200.00	74
325	323.00	44

Filtering effectiveness is measured in microns.

- Fig. 41 -

Adequate filtering for a hydraulic system normally requires the filter to remove all contamination greater than about 25 microns, and nominally (this actually means about 95% of the time) all larger than 10 microns.

There are three types of filters used in aircraft hydraulic systems. Surface filters trap the contamination on the surface of the element. Sintered metal*, a porous material made up of extremely tiny balls of metal fused together, is one of the more popular types of surface filtration. These filters usually have a bypass valve which opens to allow the fluid to bypass the element if it should clog.

Micronic filters* are made up of specially treated cellulose paper elements pleated to provide more area. Filters of this type are often installed in the return line into the reservoir where the pressure drop is low.

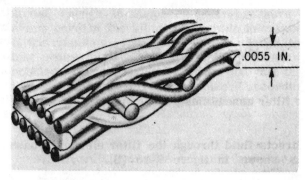

A wire mesh filter element such as this will remove about 95% of all particles larger than 5 to 10 microns.

- Fig. 43 -

Edge filters, often called Cuno filters*, are composed of stacks of thin metal discs with scrapers between them. All of the fluid flows between the discs and contaminants are stopped on the edge.

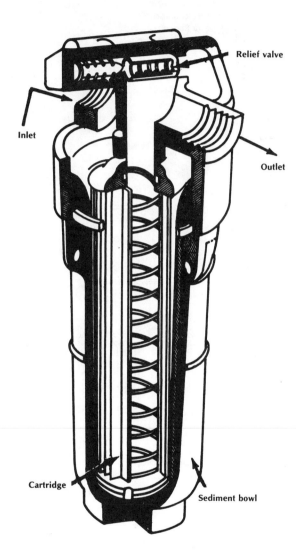

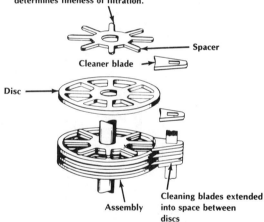

Cuno filters collect contaminants on their outer edge until they are scraped into a receptical by the cleaner blades.

- Fig. 44 -

These filters are cleaned by turning the shaft which rotates the discs and scrapes the contamination from between them into the outer housing where it can be removed by draining.

Micronic filters use a specially treated pleated paper filter element.

- Fig. 42 -

A filter similar to the one with the paper element has a stainless steel wire mesh, such as shown in Fig. 43. This wire will retain about 95% of all of the contamination larger than 5 to 10 microns.

QUESTION:

39. What is the purpose of the bypass valve in a hydraulic filter?

F. FLUID LINES

1. Rigid Lines

Most rigid lines for hydraulic or pneumatic systems are made of 5052-0 aluminum alloy. This metal is easy to form and has sufficient strength for almost all aircraft hydraulic system installations. If additional strength is needed, for instance, for higher pressure systems, or for oxygen installation, stainless steel tubing may be used.

There are two methods of attaching fittings to rigid tubing; flaring the tubing, or using flareless fittings in which a ferrule is preset so it will slightly bite into the tube to prevent its slipping out of the fitting. Aircraft fittings use a 37 degree flare, as distinguished from automotive fittings which are 45 degrees. A double flare, Fig. 45, may be used on soft aluminum alloy tubing 3/8'' in diameter and under, and a single flare, Fig. 46, may be used on all sizes.

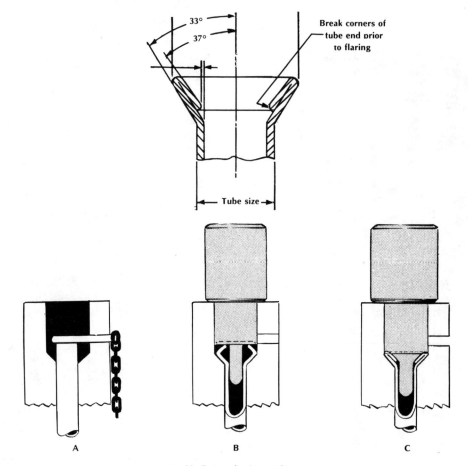

Double flaring aluminum tubing
A - Position tubing against stop;   B - Form initial upset;   C - Complete flare.

- Fig. 45 -

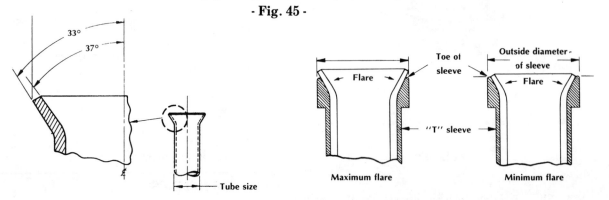

Single flare of aluminum tubing

A - Proper flare angle for AN fittings;   B - Maximum and minimum flare are measured relative to the sleeve.

- Fig. 46 -

The use of the correct procedure in making a flare is of the utmost importance. The tube must be cut perfectly square using a tubing cutter or fine tooth hacksaw. If a tubing cutter such as Fig. 47 is used, feed the cutting blade into the tube as

Parker

Typical tubing cutter for small diameter aluminum or copper tubing.

- Fig. 47 -

it is rotated around the tube. After it is cut, burr the inside edge with a special burring tool or a sharp scraper. Smooth the ends and the outside of the tube with a fine file or abrasive cloth, working **around** rather than across the end.

A number of flaring tools are available, but the one shown in Fig. 48 will single-flare tubing from 1/8'' through 3/4'' outside diameter.

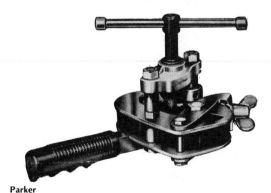

Parker

A single flare tool such as this will flare tubing from 1/8'' to 3/4'' OD.

- Fig. 48 -

Insert the end of the tubes into the tool, adjust it against the stop, and clamp it. Lubricate the cone and flare the tube until it is the correct size. Fig. 46-B illustrates the minimum and maximum amount of flare. The outside edge of the flare **must stick up above the top of the sleeve,** but its outside diameter must be no larger than that of the sleeve.

Be sure there are no chips or burrs in the tube on the flaring cone, as they will embed in the flare and weaken it. If all of the nicks and scratches are not polished out of the end of the tube, there is a good possibility the flare will split.

Flareless tubing fittings such as Fig. 49 are installed by first cutting and polishing the end of the tube as was done in the flaring process.

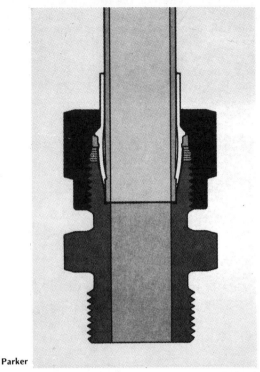

Parker

Flareless fittings are held onto the tubing by the ferrule biting into the outside of the tube near the end.

- Fig. 49 -

Slip a nut and ferrule over the end of the tube and insert it into a presetting tool, Fig. 50-A, making sure the end of the tube bottoms out on the shoulder of the tool.

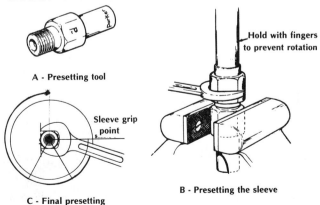

Presetting a flareless fitting
A - Presetting tool; B - Insert the tube into the presetting tool with the ferrule and nut in place. Screw the nut onto the tool until opposition is felt; C - Screw the nut down onto the tool 1-3/4 turn to crimp or preset the ferrule.

- Fig. 50 -

Lubricate the ferrule and the threads and screw the nut down finger-tight. Hold the tube against the shoulder of the tool and screw the nut down 1-3/4 turn. This presets or crimps the ferrule onto the tube. There should be a uniform ridge of metal raised above the tube surface at least 50% of the thickness of the front edge of the ferrule, and the ferrule should be slightly bowed. The ferrule may be rotated on the tube, but there must be no back and forth movement along the tube.

Almost all tubing used for aircraft plumbing has thin walls, and special care must be exercised in bending it. Minimum-bend radii must be observed, as shown in Fig. 51.

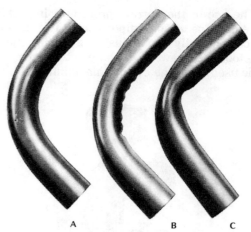

A - Good bend; B - Wrinkled bend; C - Kinked bend
Examples of good and bad bends
- Fig. 52 -

It is allowable to decrease the outside diameter of a tube to no more than 75% of its original dimension in the flattest part of the bend, so large tubing is usually bent on a production tube bender, Fig. 53, and small tubing may be bent with a hand bender similar to that in Fig. 54.

QUESTIONS:

40. Of what material are most rigid fluid lines made?
41. What flare angle is used for aircraft fluid line fittings?
42. How large should the flare be on a piece of rigid tubing?
43. Tell three ways to determine if a flareless ferrule is properly preset.
44. If a piece of 3/4 OD tubing is flattened to 5/8 inch in the flattest part of the bend, is the bend acceptable?

Tube O.D. [inches]	Minimum bend radii [inches]	
	Alum. alloy 1100-H14, 5052-0	Steel
1/8	3/8	
3/16	7/16	21/32
1/4	9/16	7/8
5/16	3/4	1-1/8
3/8	15/16	1-5/16
1/2	1-1/4	1-3/4
5/8	1-1/2	2-3/16
3/4	1-3/4	2-5/8
1	3	3-1/2
1-1/4	3-3/4	4-3/8
1-1/2	5	5-1/4
1-3/4	7	6-1/8
2	8	7

Minimum bend radii for rigid tubing

- Fig. 51 -

Fig. 52-A shows a good bend, 52-B, a wrinkled bend, and 52-C, a kinked bend.

Parker

Production tubing bender

- Fig. 53 -

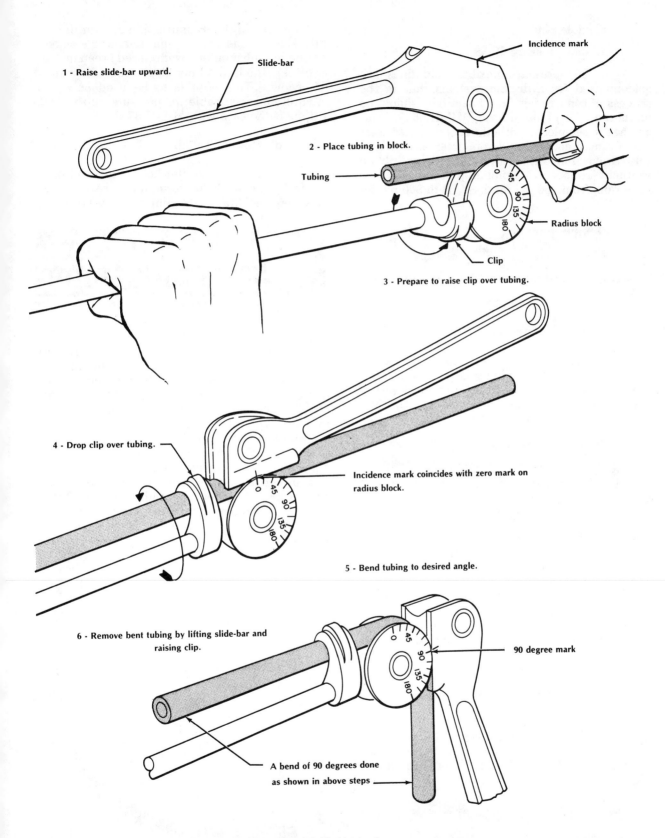

- Fig. 54 -

2. Flexible Lines

a. Low-pressure Hose

Low-pressure flexible fluid lines are seldom used for hydraulic systems, but in the process of considering fluid lines, they should be discussed. This type of hose, MIL-H-5593, Fig. 55, has a seamless synthetic rubber inner liner and a single cotton braid reinforcement. All of this is covered with either smooth or ribbed synthetic rubber. Maximum pressure for this hose runs from 150 psi for 3/8" ID to 300 psi for 1/8" ID.

Low-pressure hose used for instrument lines, but not for hydraulic systems.

- Fig. 55 -

b. Medium-pressure Hose

MIL-H-8794 hose, Fig. 56, has a smooth synthetic rubber inner liner, covered with a cotton braid. This, in turn, is covered with a single layer of steel wire braid, and over this is a rough, oil-resistant cotton braid.

MIL-H-8794 hose has one steel braid and an outer cover of rough cotton braid.

- Fig. 56 -

The operating pressure for a hose varies with its size; the smaller the hose, the higher the allowable operating pressure. Generally, MIL-H-8794 hose is used in systems operating at about 1500 psi. All flexible hoses used in aircraft fluid power systems have a lay-line, a yellow painted stripe, along its length to allow the technician to see at a glance whether or not the hose is twisted. In installing this hose, the lay line should not twist around the hose.

c. High-pressure Hose

MIL-H-8788 hose has a smooth synthetic rubber inner liner, two high-tensile carbon steel braid reinforcements, a fabric braid, and a smooth black synthetic rubber outer cover, Fig. 57.

MIL-H-8788 has two steel wire braids for reinforcement and is covered with a smooth outer cover.

- Fig. 57 -

Another high-pressure hose, similar to MIL-H-8788, has a butyl inner liner and a smooth synthetic rubber outer cover colored green instead of black. The lay and markings are white instead of yellow. This hose is to be used only with Skydrol and is suitable for pressures up to 3000 psi, as is the case with MIL-H-8788.

d. Hose of Teflon

The liner for this hose is made of tetrafluorethylene, or Teflon resin, and is covered with stainless steel braid. Medium-pressure hose, Fig. 58, is covered with one stainless steel braid, and

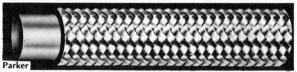

Medium-pressure hose of Teflon has a Teflon liner and a single stainless steel braid for reinforcement.

- Fig. 58 -

high-pressure hose has two. Teflon has very desirable operating characteristics and may be used in fuel, lubricant, hydraulic, and pneumatic systems in modern aircraft. This hose has one characteristic, though, that the A&P must be aware of to get the best service from it. The inner liner for this hose is extruded and will pre-form or "take a set" after it has been used with high-temperature or high-pressure fluids. After hose of Teflon has been used, it should not be bent or have any of its bends straightened out. When this tubing is removed from the airplane, it should be supported in the shape it had at the time it was installed.

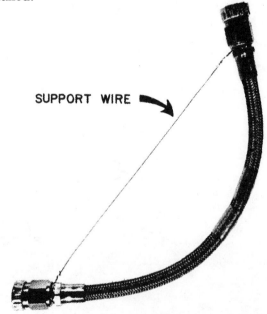

Hose of Teflon, once it has been used, takes a set, and if removed from the airplane it must be supported in the same shape it had while it was installed.

- Fig. 59 -

QUESTIONS:

45. What is the appearance of the outside of a piece of low-pressure flexible hose?

46. What is the appearance of MIL-H-8794 medium-pressure flexible hose?

47. What is the purpose of the lay-line painted along the outside of a flexible hose?

48. What is the difference in a piece of high-pressure hose with a black outer cover and one with a green outer cover?

49. What precaution should be taken when a piece of Teflon hose is temporarily removed from a hydraulic system?

G. FLUID LINE FITTINGS

1. Pipe Fittings

Some castings use National Taper Pipe Fittings to attach fluid lines, Fig. 60.

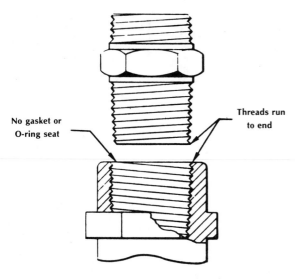

Nominal pipe size, inches	1/8	1/4	3/8	1/2	3/4	1
Threads per inch	27	18	18	14	14	11-1/2

National pipe tapered threads use no gasket or O-ring to effect their seal, but they should have a SMALL amount of thread lubricant applied to the second thread from the end to prevent galling.

- Fig. 60 -

These fittings are tapered about 1/16 inch to the inch and when they are installed the first thread should be inserted into the hole and thread lubricant applied sparingly to the second thread. Then the fitting screwed in as required. The lubricant will squeeze out between the threads and prevent their galling*, and yet there will not be enough lubricant to contaminate the system.

The size designation for tapered pipe fittings is somewhat confusing. For example, the commonly used 1/8" pipe fitting to which a 1/4 inch tube attaches does not look like 1/8 inch any way you measure it. It gets its dimensions from a standard iron pipe having an eighth inch **inside** diameter. Actually, a 1/8" NPT fitting screws into a hole about 3/8" in diameter.

2. AN Flared Fittings

Flare fittings do not depend on any type of sealant to effect a good seal; they depend rather on a good fit between the flare cone and the flare of the tube, Fig. 61.

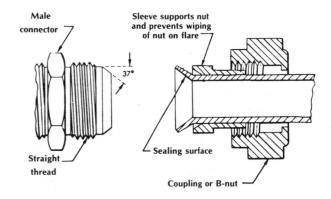

Flare type fittings depend on the contact between the flare cone and the inside of the flared tube for the seal.

- Fig. 61 -

One word of caution to the A&P: there have been two types of flare fittings used in aircraft -- the AN fitting, now commonly used, and the AC, Fig. 62.

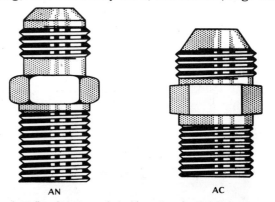

AN and AC flare fittings may look alike at first glance, but they are not interchangeable. Notice that the AC fitting has no shoulder between the flare cone and the first thread.

- Fig. 62 -

At first glance these fittings are similar, but careful observation shows that they are not interchangeable. The AN fitting has a slight shoulder between the first thread and the base of the flare cone, while the threads on the AC fitting start at the flare cone. The threads on the AC are finer than those of the AN, and, almost always, the aluminum alloy AN fittings are dyed blue while the AC aluminum fittings are yellow or gray. The nut that fits an AN fitting is noticeably longer than that used with an AC fitting. Fig. 63 shows some of the more commonly used AN flare fittings. The dash number of these fittings designates the outside diameter of the tube they

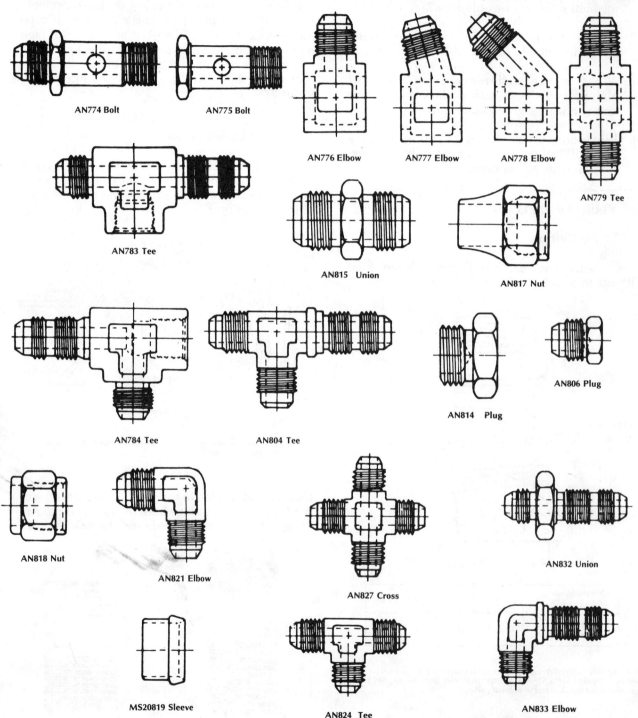

Typical AN and MS fittings for aircraft tubing installations.

- Fig. 63 -

fit. For example, an AN 815-6D union will be used to connect two 3/8" (6/16) OD tubes.

Fittings such as the AN 833 elbow are used as either positioning type fittings screwed into castings, such as pumps and valves; or as fittings to carry fluid lines through a bulkhead. Fig. 64 shows the proper way to install this fitting through a bulkhead.

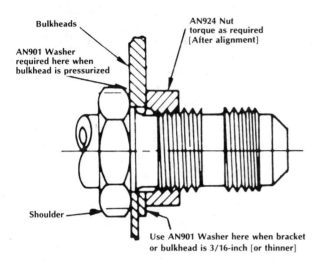

Installation of a bulkhead fitting

- Fig. 64 -

An AN 901 washer is used to back up the shoulder of the fitting and another AN 901 can be placed under the nut if the bulkhead is thin. An AN 924 nut is used to hold the fitting in the bulkhead. When this fitting is used in a casting, first screw an AN 6289 nut onto the top set of threads and work an MS9058 Teflon ring up into the counterbore of the nut, Fig. 65.

Carefully slip the proper size MS28778 O-ring over the threads, after covering them with aluminum foil, paper, or plastic to prevent their cutting the ring. Turn the nut down to push the O-ring against the lower threads and screw the fitting into the housing until the ring contacts the boss. Hold the nut with a wrench and turn the fitting in 1-1/2 turns more. This positions the O-ring in the center of the nonthreaded portion of the fitting. Now you have one full turn more in which to position the fitting so that it will line up with the line or hose you are connecting. Hold the fitting in proper alignment and torque the nut as specified in the appropriate service manual.

3. MS Flareless Fittings

There is a full line of flareless-type fittings available to be used with the crimped-on ferrule and nut. The inside of the fitting has a smooth counterbore into which the end of the tube fits, Fig. 66.

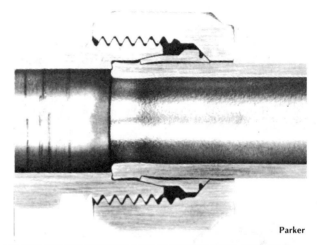

A properly preset and tightened flareless fitting will form its seal between the fitting and the ferrule.

- Fig. 66 -

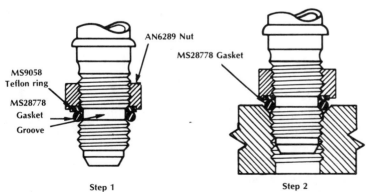
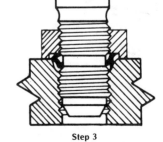

Installation of flared tube fitting in an AND10050 boss in a casting.

- Fig. 65 -

You will notice that the end of the tube bears on an angle in the fitting; this prevents the tube collapsing when torque is applied to the nut. The taper at the mouth of the fitting provides the seal between the fitting and the ferrule, and seal between the ferrule and the tube is provided by the "bite" of the ferrule into the tube.

One of the most important considerations in using a flareless-type fitting is that it not be overtorqued. When assembling a fitting of this type, be sure the ferrule is properly preset on the tube and the tube inserted straight into the fitting. Screw the nut down finger-tight, and then tighten it with a wrench one sixth of a turn (one hex); or, at the very most, one third of a turn, Fig. 67. If the fitting leaks, rather than attempt to fix it by applying more torque, disassemble it and find out what the trouble is. It is usually a damaged fitting or contamination between the ferrule and the fitting.

Fig. 68 shows a few of the many flareless tube fittings available.

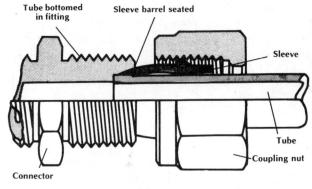

Initial alignment

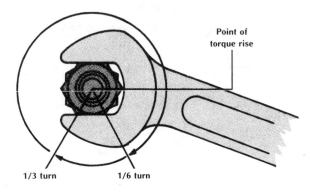

For installation locate point of sharp torque rise - then apply 1/6 to 1/3 turn

When a flareless fitting is tightened, the nut should be run down on the fitting until opposition is felt, and then the nut tightened from 1/6 to a maximum of 1/3 turn to seal.

- Fig. 67 -

QUESTIONS:

50. How much thread lubricant should be used when screwing a tapered pipe thread into a casting?

51. What is one of the easiest ways to tell the difference between an AC and an AN flare tube fitting?

52. How tight should MS flareless fittings be tightened?

MS21921 Nut MS21922 Sleeve MS21904 Elbow MS21905 Tee MS21906 Cross MS21902 Union MS21900 Adapter MS21916 Reducer

MS21913 Plug   MS21914 Cap assembly   MS21915 Bushing   MS21907 Elbow   MS21908 Elbow   MS21903 Union   MS21901 Adapter

MS 21900 −4 D
- Material — aluminum alloy
- Size of fitting in 16ths inch — 4/16 inch
- Design part number — adapter, flareless tube to AN flared tube
- Prefix — Military Specification

Typical flareless tube fittings

- Fig. 68 -

H. FLUID LINE INSTALLATION

Not only must the proper fluid lines be installed in an aircraft, they must be installed properly. Here are a few basic rules regarding the installation of fluid lines:

1. Rigid Lines

After a replacement line has been formed into the proper shape and all the bends inspected to be sure they have not collapsed, kinked, or wrinkled, the line is laid in place and checked to be sure that the tube aligns with the fittings at each end -- straight into it, with a slight pressure on the fitting.

a. **No tube, regardless of how short, should be installed unless there is at least one bend in it.** This will provide for vibration and for the inevitable expansion and contraction as the airplane changes its temperature.

b. There should never be an attempt to pull a tube up to the fitting with the nut. Doing this will place a strain on the flare or the pre-set bite and vibration can easily cause the tube to fail.

c. Where a fluid line is brought through a bulkhead, if it is not carried through with a bulkhead fitting, it must be supported with bonded cushion clamps such as Fig. 69 and centered in the hole in such a way that there is protection against chafing.

d. All fluid lines should be run below electrical wire bundles so there is no possibility of fluid dripping onto the wire.

e. All fluid lines should be identified at each end and at least once in each compartment with color coded tape to identify its contents.

f. Support clamps should be placed no farther apart than:

1/8" tubing	every 9"
1/4" tubing	every 12"
3/8" tubing	every 16"

These clamps should be placed as near the bend as possible so there will be little overhang.

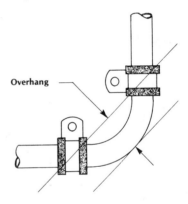

- Fig. 70 -

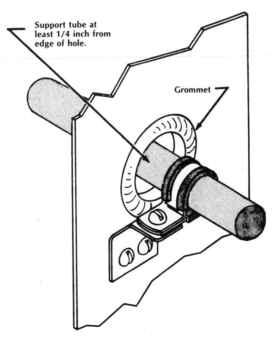

-Fig. 69 -

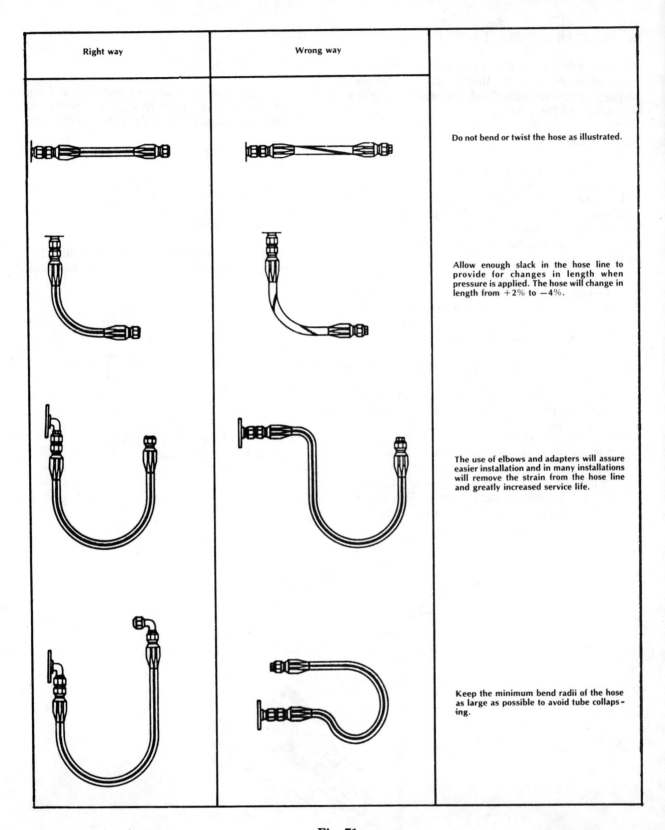

- Fig. 71 -

2. Flexible Lines

Any time there is relative movement between the two ends of a fluid line, there should be a section of flexible hose installed.

a. The lay line along a flexible hose should never spiral. This would indicate the hose was twisted and had a built-in strain. Pressure surges in a twisted line can cause failure.

b. Always use a fitting that allows the hose to approach it without any bends near its end. Elbows are available in both 90° and 45° angles, so this can be readily accomplished.

c. Never attempt to pull a hose up to its fitting by the nut. When pressure is applied to a hose, it will tend to expand its diameter and its length will shorten. Allow at least 2 to 4% slack in the line.

d. Use the proper size cushion clamp to support the hose any time it goes through a bulkhead, or any place where vibration may place a twisting force on the fitting.

e. The liner of hose of Teflon is extruded and has ample strength for applications in which there is no twist, but it is susceptible to failure if it is twisted or is bent with too small a bend radius.

f. Be sure to observe the minimum bend radius of all flexible hose. For MIL-H-8788 hose the following are minimum for each size indicated:

Hose size	Minimum Bend Radius
— 4	3.0 inches
— 6	5.0 inches
— 8	5.75 inches
—10	6.5 inches

If the hose will be subjected to flexing, this radius must be increased.

g. It is possible to make up high-pressure hose if your shop is equipped with the proper tools, but the extremely critical nature of high-pressure fluid lines makes it advisable for almost all installations to buy the proper fluid line made up and sold by the manufacturer's part number. This will assure you that the line is constructed of the proper material and has been tested according to the procedure required by the manufacturer.

h. Before installing any fluid line, be sure to blow it out with compressed air to remove any obstructions or particles that may have been left in the process of manufacture or which may have been allowed to enter while the hose has been in storage. Before a line is stored, it should have both ends capped against the entry of any contamination.

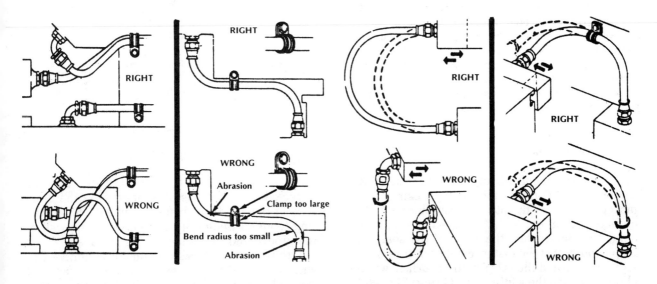

Proper clamping of flexible hoses prevents damage at the fitting.

I. HIGH-PRESSURE SEALS

Seals are used throughout hydraulic and pneumatic systems to minimize internal leakage and the loss of system pressure. There are two types of seals in use: gaskets, where there is no relative motion between the surfaces, and packings, where relative motion does exist.

1. Chevron Seals

There are many different kinds of seals used in aircraft applications, ranging from flat paper gaskets up through complex, multi-component packings. V-ring packings or chevron seals, Fig. 72, have found extensive use in the past.

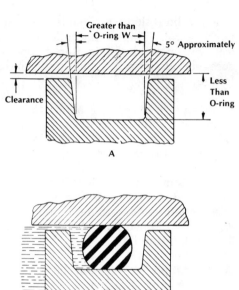

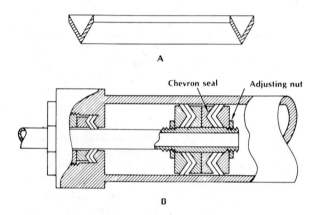

A chevron seal is a single direction seal held against the inside bore of the actuator by a spreader and an adjusting nut.

- Fig. 72 -

Fig. 72-A shows a cross section of a chevron seal, a single direction seal with the pressure on the side of the lip. Chevrons, a type of compression seal, are usually installed either in pairs or in larger stacks and require a metal backup ring and a spreader. The amount of spread of the seal is determined by the tightness of the adjusting nuts, Fig. 72-B.

2. O-ring Seals

Most modern hydraulic and pneumatic systems use O-rings for both packings and gaskets. O-rings fit into grooves in one of the surfaces being sealed. The groove should be about 10% wider than the width of the seal, and deep enough that the distance between the bottom of the groove and the other mating surface will be a little less than the width of the O-ring, Fig. 73-A. This provides the squeeze necessary to seal under conditions of zero pressure. Fig. 73-B shows the proper squeeze of an O-ring.

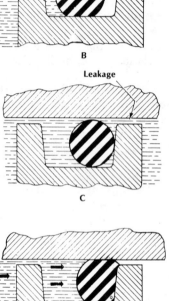

A - The groove must be wider than the O-ring, but not as deep as the ring is wide.   B - The O-ring should be pinched between the piston and the cylinder walls.   C - If there is no pinch, the O-ring will not seal.   D - Pressure on a properly fitted O-ring attempts to drive it into the opening between the piston and cylinder wall, improving the seal.

The O-ring as a hydraulic seal

- Fig. 73 -

Fig. 73-C illustrates the leakage that may be expected when there is no squeeze. As the pressure of the fluid increases, as in Fig. 73-D, the ring tends to wedge in tight between the wall of the groove and the other mating surface.

The mating surfaces should be chamfered in such a way that the squeeze will be applied gradually to the O-ring as the two are mated together, Fig. 74-A. If there is no chamfer, there is danger of damaging the ring during assembly, Fig. 74-B.

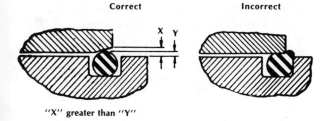

"X" greater than "Y"

For proper installation the opening of a cylinder using O-ring seals must be tapered by an amount greater than the O-ring sticks above the outside diameter of the piston.

- Fig. 74 -

An O-ring of the appropriate size can withstand pressures up to about 1500 psi without distortion, but beyond this, there is a tendency for the ring to extrude into the groove between the two mating surfaces. To prevent this, an anti-extrusion or backup ring is used.

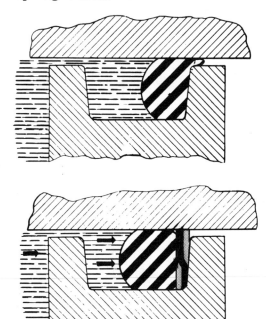

A - Pressure on an O-ring will tend to extrude it between the piston and cylinder wall.
B - An anti-extrusion ring made of Teflon or leather on the side of the O-ring away from the pressure will prevent damaging extrusion.

- Fig. 75 -

There are two types of anti-extrusion rings in use, one made of leather and the other of Teflon. Leather rings are installed in such a way that the hair side (smooth side) of the ring is against the O-ring. When installing a leather backup ring, soften it by soaking it in the fluid the ring will be used with. For pressures higher than 1500 psi, a Teflon ring is used. The ends of the spiral of the Teflon rings are scarfed, and it is possible for the ring to spiral in such a direction that the scarfs will be on the wrong side, and the ring will be damaged. Fig. 76-A is the proper way for the ring to spiral; Fig. 76-B, the improper way; and Fig. 76-C, the appearance of the ring after pressure has been applied and the ring has taken its set.

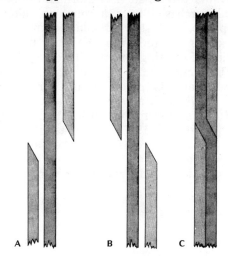

A - Proper spiral of a Teflon backup ring
B - Improper spiral
C - Teflon backup ring after it has been used and taken its set

- Fig. 76 -

The material of which the O-ring is made is dictated by the fluid of the system, and the rings are identified by colored marks on the ring:

Color	Use
Blue dot or stripe	Air or MIL-H-5606 hydraulic fluid
Red dot or stripe	Fuel
Yellow dot	Synthetic engine oil
White stripe	Petroleum-base engine oil or lubricant
White dot preceding usage mark	Nonstandard ring for use as coded
Green dash	Skydrol hydraulic fluid

Color code identification for O-rings

- Table 1 -

There is perhaps no other component as small as a hydraulic seal upon which so much importance can be placed. Seals may look alike, and it is highly probable that the wrong seal may be installed and appear to work. The material of the seal, its age, and hardness, all are important when making the proper replacement. The rule for replacing seals in a hydraulic system is to use only the **specific part number** of the seal required by the manufacturer's service information. Seals should be purchased from a reputable aircraft parts supplier, and they should be in individual packages marked with the part number, composition of the ring, manufacturer, and cure date. The cure date is the date of ring manufacture and is given in quarters. For example, 2Q75 indicates that the ring was manufactured some time in April, May, or June (second quarter) of 1975. Nominally, rubber goods are not considered fresh if they are more than 24 months old.

Of real practical consideration is the importance of buying hydraulic seals from a reputable supplier, as it is possible for out-of-date rings to be repackaged and stamped with a fresh date. The ring could be installed in good faith by an A&P and still fail because of deterioration. Yet, the mechanic is liable for the failure because it is only when an improper part is installed in an airplane that a violation of regulation is committed.

When installing O-rings, extreme care must be taken that the ring is not nicked or damaged by either sharp edges of threads or by the tool. Fig. 77 shows some of the special O-ring tools you can make. These may be made of brass rod and polished so there will be no sharp edges to nick the seal.

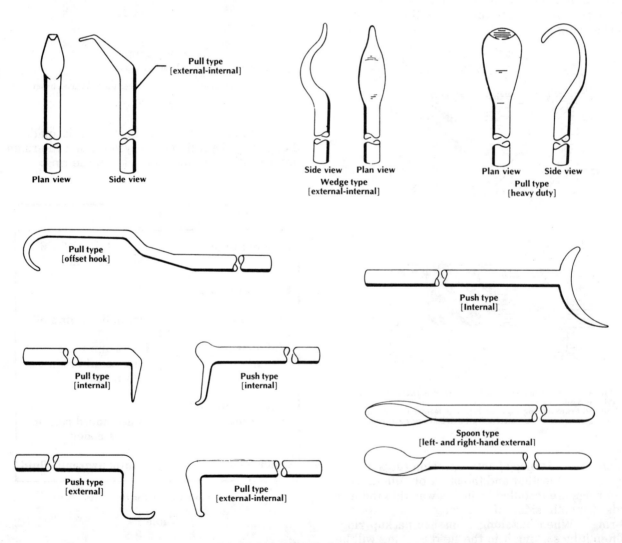

Typcal O-ring installation and removal tools.

- **Fig. 77** -

Fig. 78 illustrates the use of the O-ring tools and the proper methods of installing and removing the rings in both internal and external grooves.

When installing an O-ring over a sharp edge,

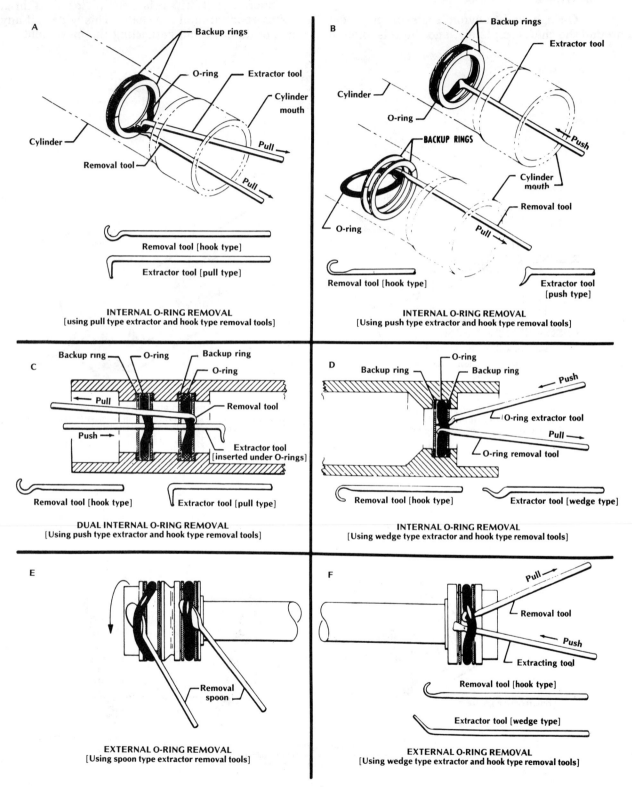

O-ring removal

- Fig. 78 -

cover the edge with paper, aluminum foil, brass shim stock, or a piece of plastic, Fig. 79.

3. Wipers

O-rings and chevron seals do not seal around the shaft completely, and there is enough leakage to lubricate the shaft. This lubricant attracts dust, and to prevent the seals being damaged when the shaft is retracted into the cylinder, a felt wiper is usually installed in a counterbore around the shaft. This wipes off any dirt or dust without restricting the movement.

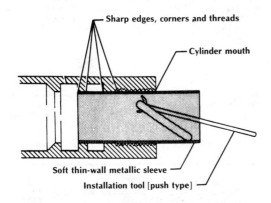

INTERNAL O-RING INSTALLATION
[Using metallic sleeve to avoid O-ring damage from sharp edges or threads, and push type installation tool]

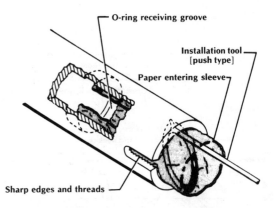

INTERNAL O-RING INSTALLATION
[Using paper entering sleeve to avoid O-ring damage from sharp edges or threads, and push type installation tool]

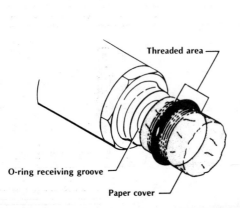

EXTERNAL O-RING INSTALLATION
[Using paper cover to avoid O-ring damage from sharp edges or threads]

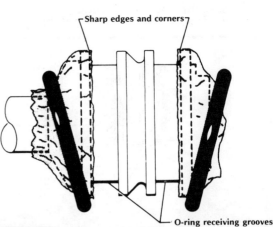

O-ring installation

- Fig. 79 -

QUESTIONS:

53. What is the difference between a gasket and a packing?

54. To which side of a chevron seal should the pressure be applied?

55. What determines the tightness with which a chevron seal fits the cylinder bore?

56. How is extrusion of an O-ring prevented in a high-pressure system?

57. When a Teflon backup ring is spiraled, should the sharp ends of the spiral be on the outside or on the inside?

58. How would an O-ring be color coded for use with Skydrol hydraulic fluid?

59. How can you know that an O-ring is the proper one for the installation?

J. ACTUATORS

1. Linear

The ultimate function of a hydraulic or pneumatic system is to convert the pressure in the fluid into work. In order to do this, there must be some movement. Linear actuators consist of a cylinder and piston. The cylinder is usually attached to the aircraft structure and the piston to the component being moved. Fig. 80.

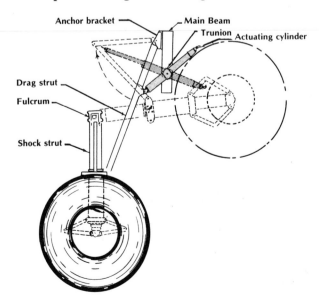

Typical retractable landing gear actuator

- Fig. 80 -

In this case, the cylinder attaches to the main wing spar through a trunion fitting.

There are three basic types of linear actuators. The single-acting actuator has the piston moved in one direction by hydraulic force and returned by a spring, Fig. 81-A, while the double-acting actuator may be either balanced, Fig. 81-B, or unbalanced, Fig. 81-C.

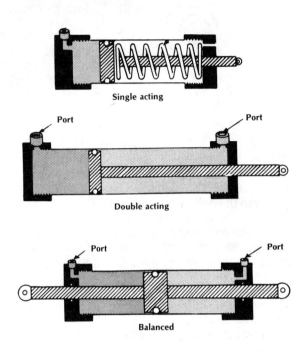

Typical linear actuators

- Fig. 81 -

Unbalanced actuators have more area on one side of the piston than on the other because of the piston rod. An example is that of Fig. 80. To raise the gear, as much force as possible is required so that the fluid pushes against the full area of the piston. The weight of the gear plus the air helps get the gear down, so a smaller amount of force is required to lower and lock the gear. The fluid is put into the end of the actuator with the rod and it pushes on that portion of the piston not taken up by the shaft.

A balanced actuator has a shaft on both sides of the piston so the area is the same on each side and the same force is developed in each direction. Balanced actuators are commonly used for hydraulic automatic pilot servo actuators.

Linear actuators may have features that adapt them to special jobs. Fig. 82 illustrates a cushioned actuator that allows the piston at the beginning of its stroke to move slowly, accelerate during the middle part of its stroke, and snub or decrease its speed at the end of the stroke.

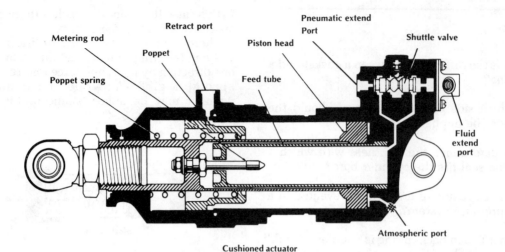

- Fig. 82 -

This is done by introducing the fluid through the shuttle for, in this case, gear extension. Fluid flows through the feed tube, around the metering pin, into the hollow piston shaft. The restriction of the fluid allows the piston to move out slowly. When the metering pin pulls all the way out of the orifice, the piston extends faster until the piston head contacts the poppet. The movement of the piston is slowed by the poppet spring, and it comes to a smooth stop at the end of its travel. Retraction of the piston is fast at first, but near the end of its stroke it is slowed by the metering pin.

Some actuators have locks to hold the piston in a retracted position until the hydraulic pressure releases them. Figs. 83 and 84 show a linear actuator used for the main landing gear. In Fig. 83, the piston is retracted and the gear extended and locked down. When the piston moves in, the retainer is pushed back, allowing the locking pin to move into position and force the balls into the groove in the piston assembly, locking the piston in place. When pressure is applied to the cylinder to raise the gear, the first thing that happens is that hydraulic pressure unlocks the piston and allows it to extend.

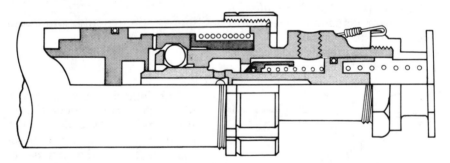

- Fig. 83 -

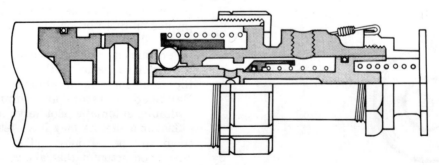

- Fig. 84 -

2. Rotary

Perhaps one of the most simple forms of rotary actuator is the rack and pinion actuator used by the single-engine Cessna aircraft for the retraction of the main landing gear. Fig. 85 shows this actuator: a simple piston with rack teeth cut in the shaft moves in and out to rotate the pinion to raise or lower the gear.

For continual rotation, hydraulic motors are used. These are similar to hydraulic pumps except for certain detail design differences. Piston motors, as in Fig. 86, have many applications on larger aircraft where it is desirable to have a considerable amount of power with good control, the ability to instantaneously reverse the direction of rotation, and no fire hazard if the motor is stalled.

Vane-type motors are also used, but instead of these being as simple as a pump, they require provision to balance the load on the shaft. This is

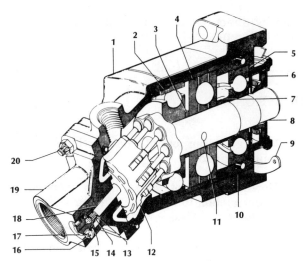

1. Housing
2. Drive shaft bearing
3. Bearing spacer
4. Thrust bearing
5. Drive shaft bearing
6. Oil seal assembly.
7. Bearing spacer.
8. Shaft and piston subassembly
9. Retaining ring
10. Bearing and oil seal retainer
11. Universal link retainer pin
12. Cylinder block
13. Spring retaining washer
14. Spring
15. Cap retaining ring
16. Retaining ring
17. Cap
18. Cylinder bearing pin
19. Valve plate
20. Valve plate mounting stud

A piston-type hydraulic motor is similar to a piston pump.

- Fig. 86 -

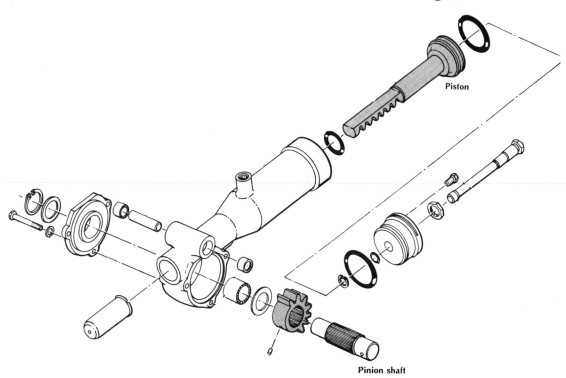

Cessna

A rack and pinion actuator changes linear movement of the piston into rotary movement of the pinion shaft.

- Fig. 85 -

done by directing some of the pressure to both sides of the motor, Fig. 87.

QUESTIONS:

60. What is the purpose of a balanced linear hydraulic actuator?

61. How is linear motion of a piston changed to rotary motion to raise the landing gear on a single-engine Cessna?

62. What are two advantages of a hydraulic motor over an electric motor?

63. What is meant by a balanced vane-type hydraulic motor?

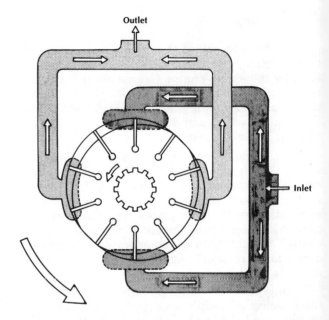

A balanced vane-type hydraulic motor directs pressure to vanes on both sides of the motor at the same time.

- Fig. 87 -

SECTION VII: AIRCRAFT LANDING GEAR

Cessna

The spring steel landing gear accepts the shock but does not actually absorb it.

- Fig. 88 -

There is perhaps no other single part of an airplane structure that takes the beating the landing gear is subjected to. A single hard landing can apply forces many times the weight of the airplane to the tires, wheels, and shock absorbing system.

A. SHOCK ABSORBERS

Not all airplanes use shock absorbers -- The popular Cessna single-engine series of airplanes does not use a shock absorber for its main gear, Fig. 88. Instead, either a steel leaf or tubular spring gear accepts the energy of the landing impact and returns it to the airplane. In a properly conducted landing, energy is returned in such a way that no rebound is caused. Another type of landing gear with a shock absorber is the bungee* shock cord gear used on the early models of Piper aircraft, Fig. 89. Elastic shock cord composed of many small strands of rubber encased in a loose weave cotton braid stretches with the landing impact and returns the energy to the airframe.

Bungee shock cord rings accept the shock but do not absorb it.

- Fig. 89 -

To absorb shock, the energy of the landing loads must be converted into some other form of energy. This is done on most modern airplanes by an oleo (oil) shock strut, Fig. 90.

The oleo strut piston is prevented turning in its cylinder by torsion links which allow in and out movement, but no rotation.

- Fig. 90 -

The wheel is attached to the piston of the oleo strut which is held in the cylinder by torsion links or scissors. These allow the piston to move in and out but not to turn. The cylinder is attached to the structure of the airplane. Fig. 91 shows the inside of the strut composed of two chambers separated by an orifice, with a tapered metering pin moving in the orifice. The strut is completely collapsed

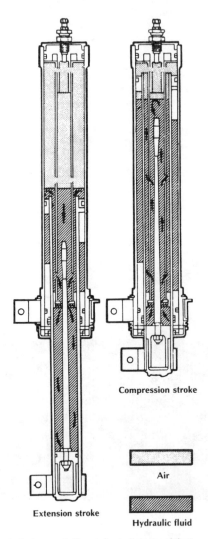

An oleo shock strut converts the mechanical energy of the impact into heat in the fluid as the fluid is forced through the small metering orifice.

- Fig. 91 -

and filled with aircraft hydraulic fluid; then compressed air or nitrogen is introduced into the strut to extend it to a specified height with the weight of the airplane on it. As the strut extends, the oil drains into the lower compartment. On touchdown, the piston is forced into the cylinder, and oil passes from the lower chamber into the upper through the metering orifice. The orifice restricts the flow, thus generating heat as the oil is forced into the upper compartment. This heat is from the energy of the landing impact. You will notice in Fig. 91 that the metering pin is tapered. As the piston is pushed farther up into the cylinder, the size of the orifice is decreased and the passage of the oil is more restricted. This provides a progressively stiffer shock strut and smoother shock absorbing action. Notice, also, that there is an enlarged area or knob at the end of the metering pin. This prevents rebound. If the pilot should bounce on landing, the strut will attempt to extend fully; but when the knob is reached, it will snub or slow down the extension.

B. WHEEL ALIGNMENT

It is important for the wheels of an airplane to be in proper alignment with the airframe. Airplanes using oleo shock struts have their wheels aligned by inserting shims between the arms of the torque links, Fig. 92.

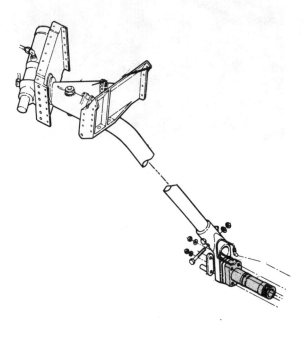

Wheel alignment on a spring steel landing gear is done with shims between the axle and the gear leg.

- Fig. 93 -

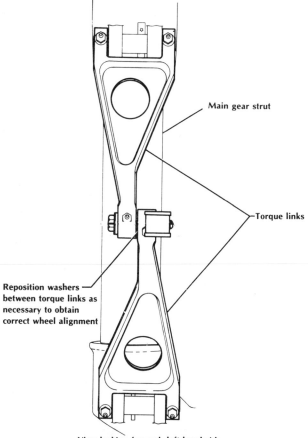

Wheel alignment on landing gear equipped with oleo shock struts is done with shims between the arms of the torque links.

- Fig. 92 -

Airplanes using spring steel landing gear have their wheel alignment changed by adding or removing shims from between the axle and the landing gear strut. Fig. 93.

C. NOSE WHEEL STEERING AND SHIMMY DAMPERS

Almost all tricycle gear airplanes have provisions for steering the airplane by controlling the nose wheel, but some of the smallest have a castering nose wheel and steering is done by independent braking of the main wheels. Some light airplanes with steerable nose wheels have a direct linkage between the rudder pedals and the nose gear; others have their nose gear steerable through a specific range, after which it breaks out of steering and is free to caster up to its limits of travel.

Because of the geometry of the nose wheel, it is possible for it to shimmy at certain speeds. To prevent this, a shimmy damper, Fig. 94, is installed between the piston and cylinder of the nose gear oleo strut. These dampers range from the most simple sealed unit of Fig. 95, through the complex rotary dampers of Fig. 96. As the nose wheel fork rotates, hydraulic fluid is forced from one compartment into the other through a small orifice. This restricts rapid movement of the gear in a shimmy but has no effect on normal steering.

Large aircraft normally have the nose gear steered by hydraulic action. This is done by

A typical nose wheel shimmy damper which can be serviced.
- Fig. 94 -

Piper

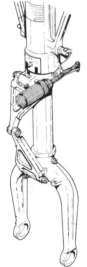

A typical sealed nose wheel shimmy damper
- Fig. 95 -

A vane-type shimmy damper used for nose wheel steering as well as damping the tendency to shimmy.
- Fig. 96 -

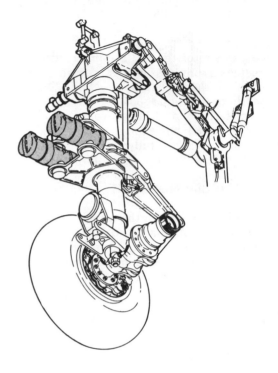

Boeing

This large transport nosewheel steering is done by two steering cylinders.
-Fig. 97 -

directing hydraulic fluid into one of the other steering cylinders, Fig. 97. Fluid from the main hydraulic system flows into the steering control valve, Fig. 98. A control wheel operated by the pilot directs pressure to one side of the nose wheel steering pistons; fluid from the opposite side is vented back into the reservoir through a pressure relief valve that holds a constant pressure on the system to snub shimmying. An accumulator in the line to the relief valve holds pressure on the system when the steering control valve is in the neutral position.

QUESTIONS:

64. What absorbs the energy of the initial landing impact in an oleo shock strut?

65. What is the purpose of the knob on the metering pin of an oleo shock strut?

66. How are the wheels on an airplane using oleo shock struts aligned?

67. How are the wheels of an airplane using a spring steel landing gear aligned?

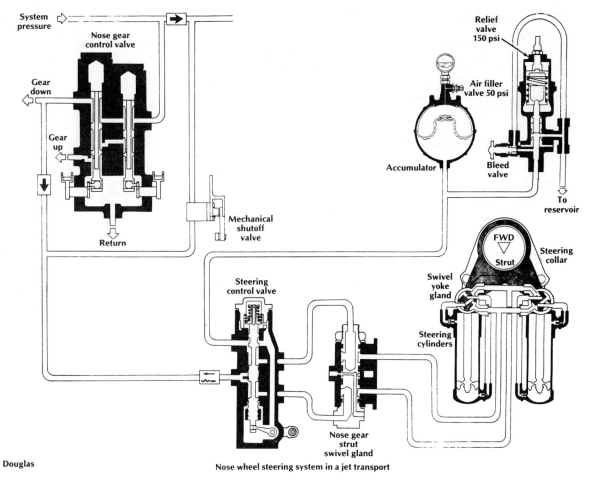

- Fig. 98 -

Nose wheel steering system in a jet transport

D. RETRACTION SYSTEMS

As the speed of aircraft becomes high enough that the parasite drag of the landing gear hanging out the airstream is greater than the induced drag caused by the added weight of a retracting system, the gear is retracted into the structure. Small aircraft use simple mechanical retraction systems, some use a hand crank to drive the retracting mechanism through a roller chain, but the most simple of all uses a direct hand lever mechanism to raise and lower the wheels. Many aircraft use electric motors to drive the gear retracting mechanism and some European-built aircraft use pneumatics. Since, however, this is a manual on hydraulics, we will concentrate on hydraulic systems.

The most simple hydraulic landing gear system uses a hydraulic power pack containing the reservoir, a reversable electric motor-driven pump, selector valve, and, in some instances, an emergency hand pump and any special valves required. Fig. 99 is a schematic of a system such as this.

To raise the landing gear, the gear selector handle is placed in the gear-up position. This starts the hydraulic pump, forcing fluid into the gear-up side of the actuating cylinders, raising the gear. The initial movement of the piston releases the landing gear down-lock, so the gear can retract. When all three gears are completely retracted, up-limit switches stop the pump. There are no mechanical up-locks, so the gear is held up by hydraulic pressure. A pressure switch starts the pump and restores pressure if it drops to a pre-determined level and any one of the wheels drops away from its up-limit switch.

To lower the gear, the selector is placed in the gear-down position, releasing the pressure on the up side of the cylinders through the power pack. The shuttle valve moves over and the gear falls down and locks. When all three gears are down and locked, the limit switches shut the pump motor off.

All retractable landing gear systems must have some means to lower the gear in the event the main extension system should fail. This simple system depends on the gear free-falling and locking into position. To actuate the emergency extension, a control on the instrument panel opens a valve between the gear-up and gear-down lines that dumps fluid from one side of the actuator to

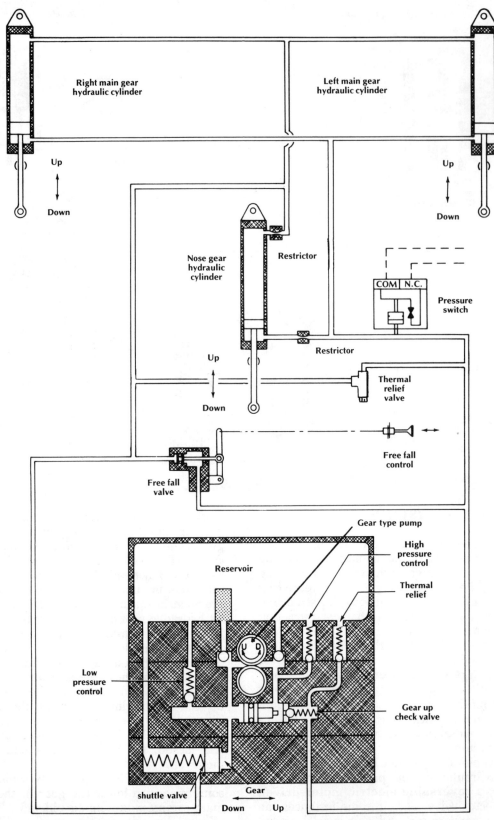

Piper

Power pack hydraulic system of a general aviation light twin aircraft.

- Fig. 99 -

the other and allows the gear to fall and lock in place.

More complex landing gear systems use compressed air or nitrogen to provide the pressure for emergency extension of the gear. In systems using this type of emergency extension, a shuttle valve is installed in the actuator where the main hydraulic pressure and the emergency air pressure meet. Fig. 100. For normal operation, fluid enters the actuator through one side of the shuttle. In the event of failure of the hydraulic system, the gear handle may be placed in the gear-down position and the emergency air supply released into the system. The shuttle valve moves over, directing compressed air into the actuator and sealing off the line to the normal hydraulic system.

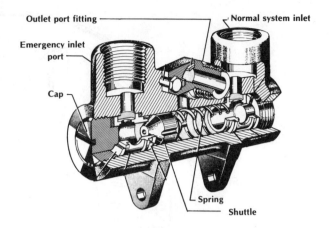

Shuttle valve

- Fig. 100 -

QUESTIONS:

68. What is used as an emergency backup for the larger hydraulically retracted landing gear?

69. What is the function of the shuttle valve in emergency landing gear extension?

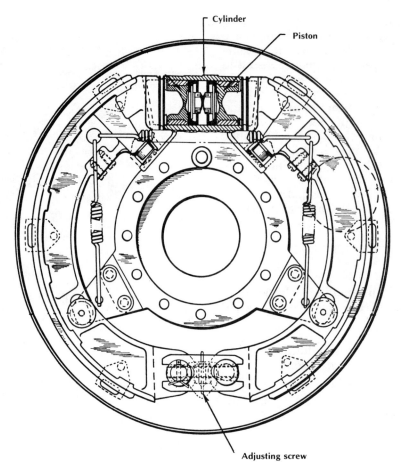

Energizing-type drum brake

- Fig. 101 -

SECTION VIII: AIRCRAFT BRAKES

A. WHEEL UNITS

Aircraft brake systems slow the airplane down by exchanging kinetic energy from the motion of the airplane into heat energy generated by the friction between the linings and the brake drum or disc. There are two basic types of brakes in use: the energizing brake, in which the weight of the airplane is used to apply the brakes, and the non-energizing brake where the weight of the airplane does not enter into the stopping action.

1. Energizing Brake

The drum-type brake, used on automobiles, is the duo-servo type; that is, the weight of the vehicle aids in the application of the brake in both the forward and rearward direction.

Some of the smaller airplanes using energizing brakes do not have duo-servo action, as only the forward motion is used to help apply the brake. Servo-action, or energizing, brakes have their shoes or linings attached to a floating back plate, Fig. 101. When the brakes are applied, the pistons in the brake cylinder move out, pushing the linings against the rotating drum. Friction attempts to rotate the linings, but they are restrained at their back end so the rotation wedges the lining against the drum. When the hydraulic pressure is released, the retracting spring pulls the linings away from the drum and releases the brakes.

2. **Nonenergizing Brakes**

   a. **Expander Tube Brakes**

   One of the early type nonenergizing brakes is that using an expander tube, Fig. 102. In this brake, hydraulic fluid from the master cylinder is directed into the synthetic rubber tube around the axle assembly, Fig. 103.

Expander tube for expander tube brakes

- **Fig. 102** -

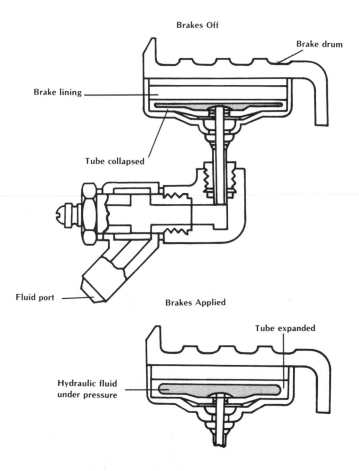

Expander tube brakes

- **Fig. 103** -

When this tube is expanded by hydraulic fluid, it pushes the lining blocks against the drum and slows the airplane. The heat generated in the lining is prevented from damaging the expander tube by thin stainless steel heat shields between each of the lining blocks.

b. Single Disc Brakes

The most popular brake for modern light aircraft is the single disc. This brake has a caliper which applies a squeezing action between the fixed linings and the rotating disc. There are two types of single disc brakes, one having the disc keyed into the wheel and free to move in and out as the brake is applied; the other has the disc rigidly attached to the wheel, and the caliper moves in and out on a couple of anchor bolts.

Fig. 104 is typical for the Goodyear single-disc brake. The disc is keyed to rotate with the wheel by hardened steel drive keys but is free to move in and out on the keys. It is prevented from rattling by disc clips. The housing is bolted to the axle and holds the two linings, one fixed in a recess in the outer portion of the housing, and the other riding in the inner side. The disc rotates between the two linings and is clamped when hydraulic fluid under pressure forces the inner lining against the

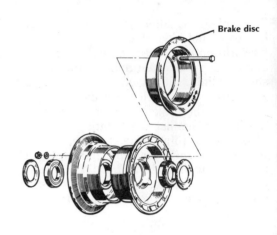

- Fig. 105 -

The disc of the Cleveland brake bolts to the wheel.

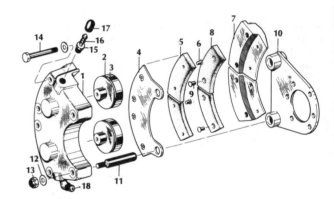

1 - Brake cylinder　　　10 - Torque plate assembly
2 - Piston　　　　　　　11 - Bolt - anchor
3 - O-ring　　　　　　　12 - Washer
4 - Pressure plate　　　13 - Nut
5 - Lining - pressure plate　14 - Bolt
6 - Rivet　　　　　　　15 - Bleeder seat
7 - Back plate　　　　　16 - Bleeder screw
8 - Lining - back plate　17 - Bleeder cap
9 - Rivet　　　　　　　18 - Elbow

- Fig. 106 -

Cleveland wheel brake assembly

disc. A piston sealed with an O-ring actually applies the push.

The Cleveland brake uses a disc solidly bolted to the inner wheel half, Fig. 105. The brake assembly, Fig. 106, consists of a torque plate bolted to the axle with two anchor bolts holding the brake cylinder, linings, and pressure plates.

When the brakes are applied, hydraulic fluid forces the pistons out and squeezes the disc between the linings.

Some of the Goodyear single-disc brakes have automatic adjusters, Fig. 107.

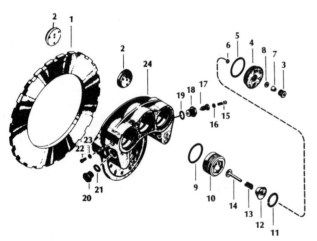

1. Brake disc　　　　　13. Brake return spring
2. Lining puck　　　　　14. Adjusting pin
3. Adjusting pin nut　　15. Bleeder screw
4. Cylinder head　　　　16. Washer
5. O-ring gasket　　　　17. Bleeder valve
6. O-ring packing　　　18. Bleeder adapter
7. Adjusting pin grip　　19. Gasket
8. Washer　　　　　　　20. Fluid inlet bushing
9. O-ring packing　　　21. Gasket
10. Piston　　　　　　　22. Screw
11. Internal retainer ring　23. Washer
12. Spring guide　　　　24. Brake housing

Goodyear single-disc brake

- Fig. 104 -

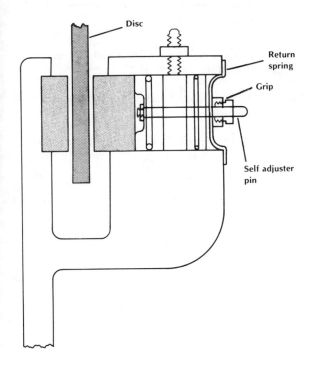

Automatic adjuster for a Goodyear brake single-disc brake

- Fig. 107 -

This feature automatically changes the amount the piston can return and in this way compensates for wear in the lining. When the brake is applied, hydraulic fluid forces the piston over and squeezes the disc between the two linings. The self-adjusting pin is pulled through the grip so that when the brake is released, the piston and lining will only move back the amount allowed by the return spring. As the lining wears, the adjusting pin is pulled into the grip and serves as an indicator of the condition of the lining.

c. Multiple Disc Brakes

It is a matter of simple physics that determines the size brake required for a given airplane. The gross weight of the airplane, the speed at the time of brake application, and the density altitude all determine the amount of heat generated in the brake. As airplane size and weight have gone up, the need for larger braking surface has increased also. This brought out, first, the thin disc multiple disc brake, and, more recently, with the advent of the jet aircraft, the segmented rotor* multiple disc brake, Fig. 108. In the brake shown here, there are five rotating discs, keyed into the wheel, and between each

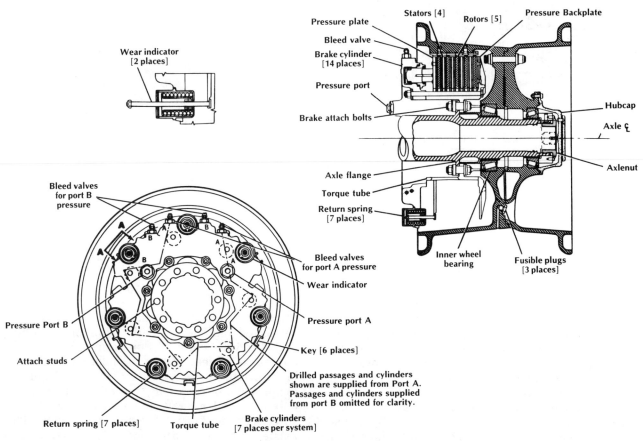

Segmented rotor brake for a jet airliner

- Fig. 108 -

65

disc there is a stator plate. Riveted to each side of these stators are wear pads made of a material which retains its friction characteristics under conditions of high temperature. A pressure plate* and a back plate* complete the stack-up. Some models of these brakes use an annular cup-type actuator to apply a force to the pressure plate to squeeze the discs together, but the one in this illustration uses a series of small cylinders arranged around the pressure plate to exert the force. Jet aircraft with two hydraulic systems to the brakes have alternate cylinders connected to each system. Automatic adjusters, Fig. 109, attach to the pressure plate and return it when the hydraulic pressure is released.

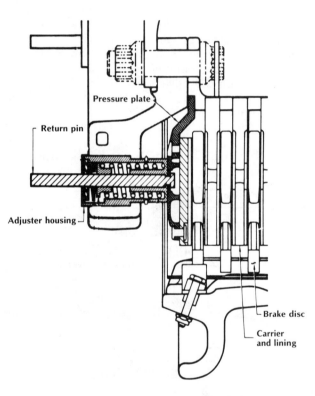

Automatic adjuster for segmented rotor brake

- Fig. 109 -

When the pressure is applied, the return pin and its retainer compresses the spring in the adjuster until the retainer bears against the housing. As the lining wears, the pin is pulled through the retainer. This is a pressed fit, so as the brake is released, the grip of the retainer is sufficient to pull the pressure plate back as much as the adjuster housing will allow. In this way, each time the brakes are applied they automatically adjust for the wear of the linings. The amount the return pin sticks out of the retainer housing is an indication of the condition of the brake linings.

QUESTIONS:

70. What is the difference between an energizing and nonenergizing brake?

71. Are single disc brakes energizing or non-energizing?

72. What is the basic principle on which automatic brake adjusters for single disc brakes work?

73. How is lining wear indicated on a brake with an automatic adjuster?

B. BRAKE ENERGIZING SYSTEMS

1. Independent Brake Master Cylinders

For years independent master cylinders have been the most commonly used pressure generating system for light aircraft brakes. The diaphragm type master cylinder, Fig. 110, is the most simple.

Scott

Diaphragm-type brake master cylinder

- Fig. 110 -

The master cylinder and brake actuator are connected together with the appropriate tubing and filled with hydraulic fluid from which all of the air has been bled. When the pilot pushes on the brake pedal, fluid is moved into the wheel cylinder to apply the brake. This type of system is useful only on small aircraft and was used with good success on the Piper Cub series of airplanes. This master cylinder is turned around and operated by a cable from a pull handle under the instrument panel on Piper Tri-Pacers. For the parking brake, a shutoff valve is located between the master cylinder and the wheel unit. The brakes are applied and the shutoff valve traps pressure in the line.

Larger aircraft require more fluid for their brakes, and there is need to vent this fluid to the atmosphere when the brakes are not applied. This prevents the brakes dragging from thermal expansion of the fluid. There are many types of vented master cylinders, but all of them have the same basic components. The Goodyear master cylinder, Fig. 111, is typical and is the one discussed here. The body of the master cylinder serves as the reservoir for the fluid and is vented to the atmosphere. The piston is attached to the rudder pedal, so when the pilot pushes on the top of the pedal, the piston is forced down into the cylinder. When the pedal is not depressed, the return spring forces the piston up so the compensating sleeve will hold the compensator valve open. Fluid from the line to the wheel unit is vented to the atmosphere. When the pedal is depressed, the piston is pushed away from the compensating sleeve, and the special O-ring and washer, the Lock-O-Seal, seals fluid in the line to the brake. The amount of pressure applied to the brake is proportional to the amount the pilot pushes. When the pedal is released, the compensator opens and vents the brake line into the reservoir, Fig. 112.

Goodyear

Goodyear vertical-type brake master cylinder

- Fig. 111 -

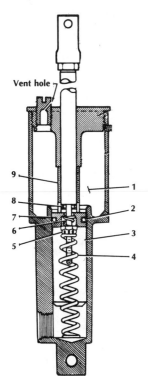

1. Reservoir
2. O-ring
3. Cylinder
4. Piston return spring
5. Nut
6. Piston spring
7. Piston
8. Lock-o-seal
9. Compensator sleeve

Cessna

Internal mechanism of Goodyear vertical master cylinder.

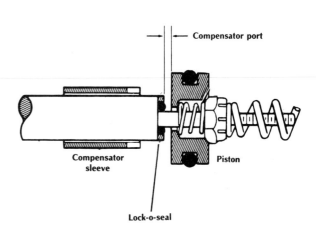

Compensator port of Goodyear vertical master cylinder

- Fig. 112 -

The parking brake for this type of master cylinder is a simple ratchet mechanism that holds the piston down in the cylinder. To apply the parking brake, the pedal is depressed and the handle pulled. This locks the piston. To release the brake, the pedal is depressed more than at the initial application so the ratchet can release.

2. Boosted Brakes

There is an airplane of a size which requires more braking force than an independent master cylinder can apply, yet does not require the complex system of a power brake; the boosted brake system is used here. In this system, the pilot applies pressure as with any independent master cylinder. If more pressure is needed than the pilot can apply, continued pushing on the pedal will introduce some hydraulic system pressure behind the piston and help the pilot apply force. This valve is installed on the brake pedal in such a way that brake application pulls on the valve. The initial movement closes the space between the poppet and the piston so fluid can be forced into the wheel unit. If there is a need for more pressure at the wheel, the pilot pushes harder on the pedal, causing the valve to straighten out. This straightening action moves the spool valve over, allowing hydraulic system pressure to get behind the piston and help apply the brakes. The spring provides regulator action and prevents pressure continuing to build up when the brake pedal is held partially depressed. As soon as the pilot releases the pedal, the spool valve moves back and relieves the system pressure back into the return manifold; the brake valve then acts as an independent master cylinder.

3. Power Brakes

a. System Operation

Almost all large aircraft use brakes operated by pressure from the main hydraulic system. This is not done by simply valving part of the pressure into the actuating units, since a brake system has special requirements that must be met. The brake application must be proportional to the force the pilot exerts on the pedals. The pilot must be able to hold the brakes partially applied without there being a build-up of pressure in the brake lines. Since these brakes are used on airplanes so large the pilot has no way of knowing when one of the wheels is locking up, there must be provisions to prevent any wheel skidding. The pressure actually supplied to the wheel unit must be lower than the pressure of the main hydraulic system, so a system of lowering or deboosting the pressure may be incorporated in the system. Since the wheels are susceptible to damage, provisions should also be made to lock off the fluid from a wheel in the event a hydraulic line is broken. Finally, there must be an emergency brake system that can actuate the wheel units in the event of a

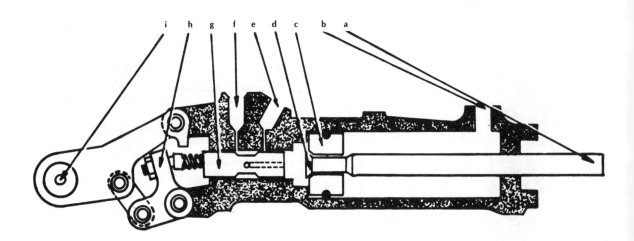

A - Pull rod;   B - Port to brake cylinder;   C - Piston;   D - Compensator poppet;   E - System pressure inlet port;   F - Port to system reservoir;   G - Spool valve;   H - Adjustment lever;   I - Attachment pivot.

Boosted brake master cylinder
- Fig. 113 -

failure of the hydraulic system.

Fig. 114 is a simplified schematic of a typical power brake system in a large jet aircraft. The brakes get their fluid from the main hydraulic system and a check valve and accumulator hold pressure for the brakes in the event of a hydraulic system failure. The pilot and copilot operate the power brake control valves through the appropriate linkages. These valves are actually regulators which provide an amount of pressure to the brake system proportional to the force the pilot applies to the pedals. Once the pressure is reached, the valve holds it as long as that amount of force is held on the pedals. In large aircraft, the pilot does not have a feel for every wheel, so an anti-skid system is installed to sense the rate of deceleration of each wheel and compare it with a maximum allowable deceleration rate. If the wheel attempts to slow down too fast, as it does at the onset of a skid, the anti-skid valve will release the pressure from that wheel back into the system return manifold.

The pressure applied by the brake control valve is too high for proper brake application, so a debooster is installed in the line between the anti-skid valve and the brake. This lowers the pressure and increases the volume of fluid supplied to the wheel units.

QUESTIONS:

74. Why do most brake master cylinders vent the fluid in the line to the atmosphere when the brake is not applied?

75. What is the difference between a boosted and a power brake?

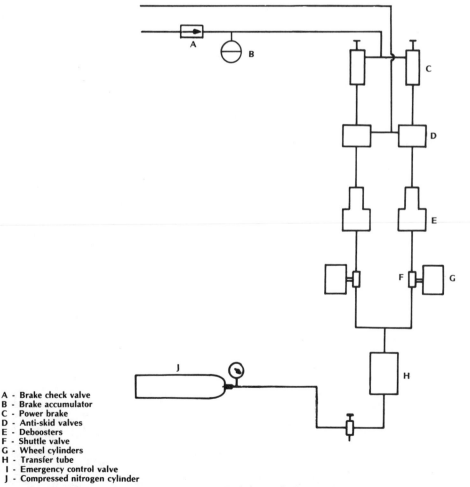

A - Brake check valve
B - Brake accumulator
C - Power brake
D - Anti-skid valves
E - Deboosters
F - Shuttle valve
G - Wheel cylinders
H - Transfer tube
I - Emergency control valve
J - Compressed nitrogen cylinder

Typical power brake system

- Fig. 114 -

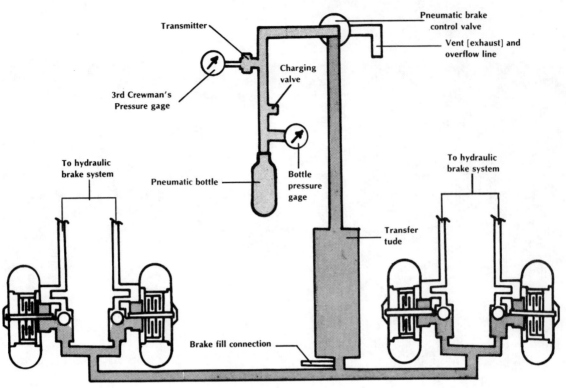

- Fig. 119 -

compressed air or nitrogen into the brake system, Fig. 119. When the pilot turns the handle, he is actually adjusting a regulator that controls air pressure to the brake. When sufficient pressure is reached in the brake line, the piston moves up against the force of a control spring and shuts off the inlet valve. The compression of the spring determines the amount of pressure supplied to the brake. When the brake handle is rotated in the direction to release the brakes, the air is exhausted overboard.

Rather than allow compressed air to enter the wheel cylinders and require the entire system to be bled of this air, the emergency air may be directed into a transfer tube where it forces hydraulic fluid into the brake system under the pressure of the emergency air.

QUESTION:

79. What type of emergency brake system does the pilot have in the event of the failure of his power brakes?

SECTION IX:
AIRCRAFT WHEELS, TIRES, AND TUBES

A. WHEELS

1. One-piece Drop Center

Fig. 120 is an illustration of a single-piece, drop center wheel. This has, because of the difficulty in changing the tire, been replaced by the more popular two-piece wheel. Tires are removed and replaced on this type of wheel by prying them over the rim.

2. Removable Flange

Some wheels have such a high flange and the tires are so stiff that single-piece wheels are impractical; so the removable rim wheel came into being. In this wheel, Fig. 121, the outer rim is removable and is held in place when the tire is inflated, with a snap ring.

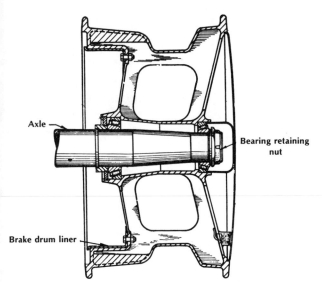

Single-piece drop center wheel
- Fig. 120 -

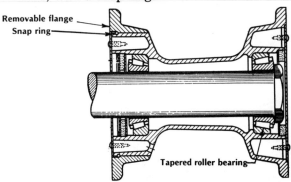

Removable rim drop center wheel
- Fig. 121 -

Great care must be exercised when inflating a tire on a wheel with a removable flange, because, if the snap ring is not properly seated, the flange can blow off and create a hazard to anyone standing nearby.

undertread to stabilize the tread for high-speed operation. It also serves as a wear indicator. When the tread is worn down to the reinforcement, it should be removed from service; but unless it is worn clear through, the tire has not been worn too much to be retreaded.

f. Tread

A layer of special rubber covers the outer circumference of the tire and serves as the wearing surface. It is grooved or dimpled to provide good braking action on the runway surfaces. Tires used on the nosewheels of some jet aircraft have a chine moulded into the tread rubber to deflect water away from the engines when landing or taking off from water-covered runways.

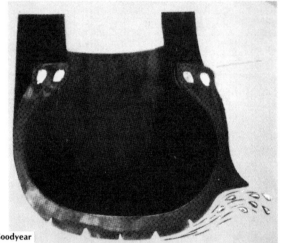

Chine-type tire for nose wheels of jet aircraft having fuselage mounted engines. This throws water away from the engine inlet.

- Fig. 125 -

g. Sidewall

This is a relatively thin rubber cover over the carcass to protect the cords from exposure and injury.

h. Liner

In tubeless tires, a layer of rubber especially compounded to resist air leakage is vulcanized to the inside of the carcass. If a tube is to be used in the tire, a thin layer of smooth rubber is used to prevent chafing the tube. Tubes should never be installed in tubeless tires because of the roughness of the inner liner.

2. Tire Types

There have been nine types of tires established by the aircraft tire and rim industry, but only three are of primary interest for civilian aviation.

a. Type III -- Low-Pressure

This type is used on most of the smaller general aviation aircraft. It has a large volume and uses low inflation pressures to provide a high degree of flotation. Type III tires have a maximum ground speed rating of 120 MPH. Its size is identified by the section width and rim diameter: for example, a 7.00-6 tire has a width of seven inches, and fits a six inch diameter rim.

b. Type VII -- Extra High Pressure

These tires are used on military and civilian jet aircraft and have ground speed ratings up to 250 miles per hour; inflation pressures for some of them are as high as 315 psi. The size of these tires is identified by their outside diameter and section width; for example, a 34 x 11 tire has a 34 inch outside diameter and a section width of eleven inches.

c. New Design

All of the new designs of aircraft tires are being dimensioned according to this category, rather than by the methods of the older types: A tire identified as 15 x 6.0-6 has an outside diameter of 15 inches, a section width of six inches, and a bead seat diameter of six inches.

QUESTIONS:

80. How are the halves on a two-piece wheel used with a tubeless tire sealed against air leakage?

81. What is the purpose of a thermal fuse in a large aircraft wheel?

82. What factors determine the ply rating required by a tire?

83. Why is it bad practice to install an inner tube in a tubeless tire?

3. Tire Inspection and Repair

a. Tread Wear

Aircraft tires take a pounding on every landing and the tread is rapidly worn off as the rough surface of the runway is used to accelerate the wheel from zero to touchdown velocity. Most aircraft tires have their tread used up long before

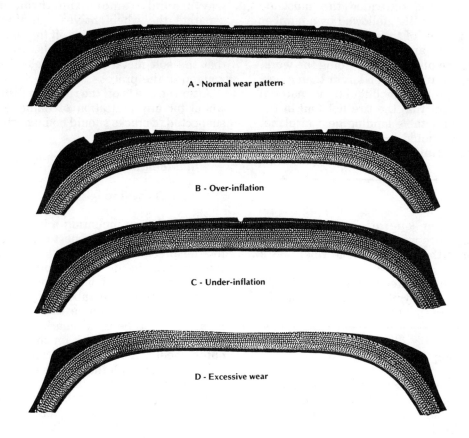

Tire tread wear is a good indicator of the tire's operating condition.

- Fig. 126 -

the carcass is worn out, so it is possible to retread them and thus extend their useful life. A careful inspection and attention to any small defect and wear indication will help get the maximum use of a tire.

Fig. 126-A is the cross section of a tire showing normal wear on the tread. This tire has been properly inflated, as is indicated by the even wear of all sections of the tread.

Fig. 126-B shows the way a tread will wear if the tire has been operated in an over-inflated condition. The center of the tire wears first and the tread ribs at the edges show less wear.

Operating a tire with too little air pressure will cause the tread ribs on the sides to wear first and the center ribs to show less wear; Fig. 126-C.

A tire should be removed from service by the time the tread reinforcement shows through. Fig. 126-D shows a tread section that has worn down to the breaker plies and is too far worn to be retreaded.

b. Retreading

Retreading can be done and effect quite a savings in the operating budget if certain precautions are observed. There are several types of retreading used on aircraft tires. Top-capping may be used when there is little shoulder wear; only the tread is removed and a new one applied.

Full capping is done by removing the old tread and applying new material which comes down several inches over the shoulder of the tire. A three-quarter retread replaces the full tread and one side of the sidewall. For this, the old sidewall is buffed down and new material is applied from the bead to the edge of the new tread.

Tires may be retreaded if the carcass is sound and there are no flat spots which extend into the carcass plies. Any of the following conditions, though, are cause for rejecting the tire:

1. Any injury to the beads or in the bead area.
2. Weather checking or ozone cracking which exposes body cords.
3. Kinked or protruding beads.
4. Ply separation.
5. Loose, damaged, or broken cords inside the tire.
6. Broken or cut cords in the sidewall or shoulder area.

Only approved repair stations may retread aircraft tires, and the retreaded tire must be identified by the letter "R" followed by a number indicating the number of times the tire has been retreaded, the month and year the retreading was done, and the name of the agency doing the work.

Since a retreaded tire is often larger than a new one, because of a greater amount of tread material used, it is important that the tire not foul in the wheel well of a retractable landing gear airplane. A retraction test should be performed after a retreaded tire is installed.

QUESTIONS:

84. What improper operating condition is indicated by a tire wearing in the center tread rib more than the shoulder tread ribs?

85. What would be indicated by the name of a tire company and the symbols R-2, 6-75 on the tread of an aircraft tire?

---

4. Demounting Tires

Before demounting a tire from an airplane wheel, all of the air should be let out by removing the core from the air valve. Be careful that the core does not blow out and injure anyone. After all of the air is out, break the bead from the rim. This can be a difficult operation if a proper tool is not used. Fig. 127 shows a form of bead breaker that applies a concentrated force to the bead and breaks it away from the rim with a minimum of effort or possibility of damaging the wheel.

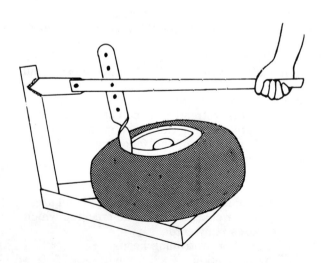

A special bead breaker is needed to break the bead away from the wheel without damaging either the wheel or the tire.

- Fig. 127 -

After the beads of both sides are broken all the way around, remove the through-bolts and separate the wheel halves. If the wheel is a one-piece unit, pry the tire off the rim, taking little bites with the proper tire tools. Be very careful that the soft metal of the wheel is not nicked or gouged in the process.

After the tire is off the rim, carefully inspect the wheel for any indication of damage. Any area suspected of cracks should be inspected with a dye penetrant.

5. Mounting Tires

a. Tube Type

When mounting a tube type tire on a two-piece wheel, first inspect the inside of the tire for the presence of any dirt or foreign objects, then dust the inside with talcum. Insert the tube and inflate it until it just rounds out. Align the balance mark of the tube, indicating the heavy point, with the mark on the tire which is at its light point. If the tube is not marked, align the valve stem with the light mark on the tire. Insert the outer wheel half into the tire, carefully guiding the valve stem through the valve hole. Slip the inner wheel half into the tire, being very careful that the tube is not pinched, install the through bolts and torque them to the values recommended by the manufacturer. Tighten them in the sequence specified in the aircraft service manual. Inflate the tube as recommended by the manufacturer, then deflate it completely. This allows the tube to seat in the tire and relieves any stresses. Reinflate the tube to the pressure recommended by the airframe manufacturer.

b. Tubeless Type

Be very sure the mounting instructions of the airframe manufacturer are followed in detail, but, in general, inspect the wheel for any indication of damage and the inside of the tire for foreign material. Lubricate the O-ring and be sure it is properly installed in the ring groove. Align the balance mark of the tire with the valve in the wheel and assemble the wheel. Torque the through-bolts as recommended by the wheel manufacturer and inflate the tire to the recommended pressure.

6. Inflating Tires

Proper inflation is the most important maintenance function to get maximum service from aircraft tires. The pressure should be checked daily with an ACCURATE pressure gage. Pressure should be checked when the tires are cool, at least two or three hours after the airplane has been flown.

When tube-type tires are installed, air is usually trapped between the tube and the tire. This gives a false pressure indication until the trapped air seeps out around the beads or around the valve stem. Tubeless tires must be checked for proper inflation after allowing a waiting period of from 12 to 24 hours, because the nylon of which the plies are made will stretch and decrease the pressure.

When a tire is installed on the airplane, and weight is put on it, the volume of the air chamber is decreased so that the pressure will be about 4% greater than with no load applied.

If a tire consistently loses air pressure, it should be carefully checked before condemning it. Tubeless tires must be **fully** inflated and allowed to sit for 12 to 24 hours and rechecked. In this period of time they will lose approximately 10% of their pressure due to the nylon stretch. Another natural cause of pressure loss is temperature change. Tire pressure will change about one psi for every 4° F. change in temperature.

Tubeless tires have vent holes in their sidewalls so that air trapped in the cords can diffuse. Air will seep from these holes throughout the life of the tire, but if the loss is more than about 5% in 24 hours, there is a possibility the inner liner has been damaged.

Inflation pressures should be checked with a gage which is known to be accurate. Dial-type gages are the best for any type of tire maintenance, and these should be periodically calibrated.

C. TUBES AND TUBE REPAIR

An aircraft tube must always be the one specified for a particular tire, as one too large will wrinkle and chafe in the tire, and one too small will stretch and lose its required strength.

When inspecting a tube out of the tire, it should never be inflated with more than just enough air to round it out. Excessive inflation will strain the splices and areas around the valve stem.

Tube damage which is repairable is usually either a damaged valve, or a cut or puncture. Wrinkling or thinning because of brake heat is usually cause for replacement of the tube and may not be repaired.

Rubber valve stems are replaced by curing a new stem and base to the outside of the tube. Procedure for valve replacement must follow the manufacturer's recommendations in detail.

Cuts and punctures may be repaired by a patch, either cemented on or cured. The cured or vulcanized patch is recommended, and its application is the same as that used for automobile tubes.

D. TIRE AND TUBE STORAGE

Tires and tubes, as with any other rubber goods, should be stored in a cool, dark area away from air currents or electrical equipment. Electric motors, fluorescent lights, battery chargers, or electric welding equipment generate ozone which rapidly ages rubber goods.

Tires should be stored vertically in racks so that there is no distortion of the tire. If it is necessary to store them on their sides, they should not be stacked more than four high unless provisions are made to prevent the bottom tires being crushed.

Tubes should be stored in their original box to protect them. If it is necessary to store a tube outside of its box, it should be slightly (less than one psi) inflated inside of its proper tire. Tubes stored in this way must be removed from the tire and carefully inspected before they are mounted on a wheel, as it is possible for foreign matter to have gotten between the tire and the tube and caused damage.

QUESTIONS:

86. What kind of inspection should be used on a wheel suspected of being cracked?

87. Does the balance mark on a tire indicate its heavy or its light point?

88. Why should an inner tube be inflated and deflated before its final inflation?

89. Why do tubeless tires often have small holes in their sidewalls?

90. Why should rubber goods not be stored near an electric motor?

91. Should tires be stored vertically or horizontally?

SECTION X: HYDRAULIC SYSTEM TROUBLESHOOTING

A. BASIC PRINCIPLES OF TROUBLESHOOTING

Regardless of what we are troubleshooting, the principle is the same: isolate the problem, and then eliminate everything in that area that is good. Whatever is left is bound to contain the trouble.

It almost goes without saying that isolation of a problem cannot be accomplished unless you know the system thoroughly. Modern hydraulic systems are complex and no troubleshooting can be logically attempted without having the schematic of the system before us.

Hydraulic systems logically divide themselves into power and actuation systems. The power system can be further divided into the main, backup, and emergency systems. If a malfunction occurs when operating on the main power system, the engine-driven pump, but actuation is normal with the hand pump, the trouble could be low fluid in the reservoir (below the engine pump outlet), or it could possibly be in the pump, itself. Always check the most simple possibility first, and eliminate it before going on. If neither the main system nor the hand pump could get the landing gear down, but it was lowered by the emergency system, the actuation system is good; the trouble will be in the power system, such as no fluid, or an unloading or relief valve stuck open.

If only one system fails to actuate and there is pressure in the system, the trouble is in the actuation system. If, for instance, the flaps fail to actuate, but the landing gear actuates normally, the trouble is most probably somewhere between the flap selector valve and the actuator.

B. TROUBLESHOOTING TIPS AND PROCEDURES

1. Pumps are fluid movers, not pressure generators. If there is no pressure indicated on the system pressure gage, yet there is a return flow into the reservoir, the trouble most probably does not lie with the pump but in the pressure control valves for the system. A regulator or relief valve is probably stuck open.

2. System pressure higher than normal could quite likely be caused by the unloading valve failing to unload the pump and a relief valve maintaining the pressure. Knowing the setting of each relief valve will help determine which valve is doing the work. When a relief valve holds the system pressure, it will normally make a buzzing noise and will get quite warm.

3. A loud hammering noise in a system having an accumulator indicates an insufficient air pre-load in the accumulator. The pump goes on the line or kicks in, and since there is no compressible fluid in the accumulator, the system pressure immediately builds up to the kick-out pressure and the pump goes off the line. This kicking in and out without any air to compress and cushion the shock causes the heavy hammering. Most systems do not have air gages on the accumulator to determine the pre-load, so, to find out how much air is in the accumulator, pump the hand pump slowly and watch the hydraulic pressure gage. It will not rise at first, but then it will suddenly jump up and, as you continue to pump, will rise slowly again. The pressure jumped at the point fluid was first forced into the accumulator where it was opposed by the air, pressurizing the hydraulic system. This indicates the amount of air pre-load.

4. Pump chattering and subsequent overheating is caused by air in the line. The most logical place for the air to enter is from a low reservoir. Another place is a leak in the suction line between the reservoir and the inlet of the pump.

5. Slow actuation of a unit is often caused by internal leakage in a valve or actuator. This also causes the pump to kick on and off the system more often than it should. If the system uses an electric pump, you will notice the ammeter shows the pump to be operating quite often when nothing is being actuated. The offending component may be heated by the fluid leaking so it can be found by feeling of all of the suspected units and checking the one that is unnaturally warm.

6. Spongy actuation is usually a sign of air in the system. Most double-acting systems are self-bleeding; so, after a component has been replaced, it should be cycled a number of times to purge all of the air in the actuator and the lines back to the reservoir. If a system will not purge normally, the manufacturer will usually have special instructions in the service manual to explain the method of removing all of the residual air.

Brakes are single-acting system and require bleeding any time the system has been entered. There are two methods of removing air from brake systems:

 a. Gravity bleeding: A clear plastic tube is attached to the bleeder plug at the wheel cylinder and the end of the tube immersed in a container of clean hydraulic fluid. The reservoir is filled, and the brake pedal depressed slowly with the bleeder open. Watch the plastic tube as you continue to slowly pump the master cylinder. Do this until fluid runs through the tube with no bubbles. You may have to fill the reservoir a time or two in this process, as a fair amount of fluid will have to be pumped through the system before all of the air is removed. When the fluid runs clear of all traces of air, close the bleeder plug and remove the tube; fill and cap the reservoir, making sure the vent is open.

 b. Pressure bleeding: This type of bleeding is usually superior to gravity bleeding since it begins at the low point and drives air out the top. Connect a hose to the bleeder plug at the wheel cylinder and attach a bleeder pot or hydraulic hand pump. Attach a clear plastic hose to a fitting in the top of the reservoir and immerse its free end in a container of clean hydraulic fluid. Open the bleeder plug and slowly force fluid through the brake, up through the reservoir and out into the container of fluid. When the fluid flows out of the reservoir with no trace of air, close the bleeder plug, and remove the hoses. Some reservoirs may be overfilled in this process and fluid must be removed down to the "Full" mark before replacing the reservoir cap. Be sure the vent is open when capping the reservoir.

7. Many systems have multiple relief valves, set at different pressures. To check these valves, screw the adjustment on all the valves down beyond the range of the highest pressure valve and adjust this valve first. Then adjust the rest of them, setting them in the order of descending pressure.

8. Sometimes a pressure gage will fluctuate rapidly with the pointer forming a blur. This is an indication of air in the gage line, or insufficient snubbing. Hold a shop rag around the back of the instrument and crack the fitting just enough to purge any air from the line, then retighten it. If this does not cure the fluctuation, the gage snubber must be replaced.

Troubleshooting hydraulic and pneumatic systems is usually a logical application of basic principles, and we as A&P technicians are charged with maintaining certificated airplanes in the condition of their original certification. This simply means that all of our work must follow the recommendations of the manufacturer. When we have a specially sticky problem, we should not hesitate to contact the manufacturer's service representatives for help. He has probably encountered the problem before and can save us time and money in getting the airplane back into productive flying.

QUESTIONS:

92. What are the two basic steps in logical troubleshooting?

93. If a relief valve is holding the system pressure, will the pressure be higher or lower than normal?

94. How can an internal leak in a hydraulic component sometimes be detected?

95. Why is pressure bleeding of the brakes usually superior to gravity bleeding?

GLOSSARY

This glossary of terms is to give a ready reference to the meaning of some of the words with which you may not be familiar. These definitions may differ from those of standard dictionaries but are more in line with shop usage.

back plate: A floating plate on which the wheel cylinder and brake shoes attach on an energizing-type brake.

area: The number of square units in a surface.

balanced actuator: A hydraulic or pneumatic actuator having the same area on both sides of the piston.

Bernoulli's principle: The basic principle of fluids in motion. When the total energy remains constant, any increase in kinetic energy in the form of velocity will result in a decrease in potential energy in the form of pressure.

bungee shock cord: A shock absorbing medium composed of numerous rubber bands enclosed in loosely woven fabric.

Butyl: The trade name of a synthetic rubber product made by the polymerization of isobutylene. It withstands such potent chemicals as Skydrol hydraulic fluid.

case pressure: A low pressure maintained inside the case of a hydraulic pump. In the event of a damaged seal, fluid will be forced out of the pump rather than allowing air to be drawn in.

circle: A closed plane figure with every point an equal distance from the center. A circle has the greatest area for its circumference of any enclosed shape.

constant: A valve used in a mathematical computation that is the same every time. For instance, the relationship between the circumference of a circle and its diameter is a constant, 3.1416 (pi).

constant displacement pump: A pump which displaces a constant amount of fluid each time it turns. The faster it turns, the more it puts out.

Cuno filter: The proprietary name of a fluid filter made up of a stack of discs separated by scraper blades. Contaminants collect on the edge of the discs and are periodically scraped out and collected in the bottom of the filter case.

double-acting actuator: A linear actuator which is moved in both directions by fluid power.

epoxy: A flexible, thermosetting resin made by the polymerization of an epoxide. It is noted for its durability and chemical resistance.

fluid: A substance, either a gas or a liquid, which flows and tends to conform to the shape of its container.

fluid power: The transmission of force by the movement of a fluid. The best examples are hydraulics and pneumatics.

force: Energy brought to bear which tends to cause a motion or change.

galling: Fretting or pulling out chunks of a surface by sliding contact with another surface or body.

gerotor pump: A form of gear pump which uses an external spur gear, inside of and driving an internal gear having one more tooth space than the drive gear. As the gears rotate, the space between two of the teeth increases, while that on the opposite side of the pump decreases.

gram: The unit of weight or mass in the metric system. One gram equals about 0.035 ounce.

hydraulics: that system of fluid power which transmits force by an incompressible fluid.

hydrostatic paradox: A condition that does not at first appear to be true, in which it can be observed that the pressure exerted by a column of liquid is dependent on its height and independent of its volume.

kilogram: One thousand grams.

kinetic energy: Energy possessed by an object because of motion.

micron: One millionth of a meter. It is normally used to denote the effectiveness of a filter.

Micronic filter: The trade name of a filter having a porous paper element.

moment: The product of the weight of an object in pounds, and the distance from the center of gravity of the object to the datum or fulcrum in inches. Moment is used in weight and balance computations and is expressed in pound-inches.

naphtha: A volatile and flammable hydrocarbon liquid used chiefly as a solvent or cleaning agent.

orifice check valve: A component in a hydraulic or pneumatic system that allows unrestricted flow in one direction, and restricted flow in the opposite direction.

Pascal's law: A basic law of fluid power which states that pressure in an enclosed container is transmitted equally and undiminished to all points of the container and acts at right angles to the enclosing walls.

piston: A sliding plug in an actuating cylinder used to convert pressure into force and then into work.

pneumatics: That system of fluid power which transmits force by the use of a compressible fluid.

polyurethane enamel: A hard, chemically resistant finish whose long flow-out time and even cure throughout give a flat surface and a glossy "wet" look. Especially suitable for seaplanes and agricultural planes because of its resistance to chemical action and abrasion.

polyvinyl chloride: A thermoplastic resin used in the manufacture of transparent tubing for electrical insulation and fluid lines which are subject to low pressures.

potential energy: That energy possessed by an object because of its position, configuration, or the chemical arrangement of its constituents.

power [P]: The time rate of doing work. Force times distance, divided by time. Electrically it is the product of voltage (E) and current (I); P = EI.

pressure: Force per unit area.

pressure plate: A heavy strong plate in a multi-disc brake which receives the force from the brake cylinders and transmits it to the discs.

rack and pinion actuator: A form of rotary, actuator where the fluid acts on a piston on which a rack of gear teeth is cut. As the piston moves, it rotates a mating pinion gear.

rectangle: A plane surface with four sides whose opposite sides are parallel and whose angles are all right angles.

segmented rotor brake: A heavy-duty multiple disc brake used on large, high-speed aircraft. Stators are keyed to the axle and contain high-temperature lining material. The rotors, keyed into the wheel, are made in segments to allow for cooling and for the large amounts of expansion encountered.

shuttle valve: A valve mounted on critical components which directs system pressure into the actuator for normal operation, but emergency fluid when the emergency system is actuated.

silicone rubber: An elastic material made from silicone elastomers. It is used with fluids which attack other natural or synthetic rubbers.

single-acting actuator: A linear hydraulic or pneumatic actuator which uses fluid power for movement in one direction, and a spring force for its return.

sintered metal: A porous material made up by fusing powdered metal under heat and pressure.

Skydrol hydraulic fluid: A synthetic, nonflammable, ester base hydraulic fluid used in modern high-temperature hydraulic systems.

square: A four sided plane figure whose sides are all the same length, whose opposite sides are parallel, and whose angles are all right angles.

static: Still; not moving.

Stoddard solvent: A petroleum product similar to naphtha, used as a solvent or cleaning agent.

Teflon: A proprietary name for a fluorocarbon resin used to make hydraulic and pneumatic seals and backup rings.

triangle: A three sided, enclosed plane figure.

turbine: A rotary device actuated by impulse or reaction of a fluid flowing through the vanes or blades arranged around a central shaft.

variable displacement pump: A pump whose output may be varied by the pressure on the system. For high-pressure applications, this is usually done by varying the stroke, either actual or effective, of a piston-type pump.

varsol: A petroleum product similar to naphtha, used as a solvent.

volume: Space occupied. It is measured in cubic units.

work: The product of force and distance.

AIRCRAFT HYDRAULICS
Answers to study questions

1. Hydraulics normally deals with force transmission using a noncompressible fluid, while pneumatics uses a compressible fluid.

2. The rectangle has twice the area.

3. Approximately 19 square inches.

4. Work.

5. 33,000.

6. 26.24 horsepower.

7. The height of the column.

8. The product of the force, and the distance between the point of application and the fulcrum.

9. Kinetic and potential.

10. Yes.

11. Alcohol.

12. Naphtha, varsol, or Stoddard solvent.

13. Trichlorethylene.

14. Refer to the aircraft service manual.

15. The system should be drained and thoroughly flushed as recommended by the airframe manufacturer.

16. Parallel.

17. Series.

18. Compressed air or nitrogen.

19. The pump will be automatically unloaded.

20. When the air is expanded for use, the temperature will drop enough to freeze the moisture and plug the system.

21. It isolates the air supply from the actuation systems and allows the actuators to be serviced without emptying the air bottles.

22. (1) Holds the supply of fluid.
    (2) Purges air from the system.
    (3) Provides room for the expansion of fluid.

23. The reservoir is pressurized.

24. Hydraulic pumps move fluid.

25. Around the outside of the gears.

26. In case the shaft seal is damaged, fluid will leak out rather than air being drawn into the pump.

27. Piston type.

28. The pump changes its output automatically as the system pressure varies.

29. Compensator spring force, and the hydraulic force on the compensator stem piston.

30. In an open center selector valve, fluid flows through the valve to unload the pump. A closed center valve has no central through-passage.

31. A sequence valve is mechanically actuated while a priority valve is actuated by hydraulic pressure.

32. (1) Rate of flow.
    (2) Volume of flow.

33. It would normally be placed in the up line, allowing full flow to get the gear up, but restricted flow to lower the gear.

34. The pump must work to hold pressure against the relief valve.

35. Kick-in = 1500 psi.
    Kick-out = 2000 psi.

36. A regulator adjusts the pressure and maintains it whether or not there is any flow in the system. A relief valve holds pressure in the system by restricting the return flow to the reservoir. A pressure reducer is a variable orifice, dropping the pressure to a component as long as there is flow.

37. Pressure can be stored by forcing fluid into a container having a movable partition and a compressible fluid or a spring which holds a force on the incompressible fluid.

38. High-pressure valve cores have the letter "H" embossed on the end of the stem.

39. In case the filter plugs, the bypass valve will open and allow a flow of fluid. It is better to have unfiltered fluid than no fluid.

40. 5052-0 aluminum alloy.

41. 37 degrees.

42. It must stick up above the top of the sleeve but be no larger than the outside diameter of the sleeve.

43. (1) The ferrule must be slightly bowed.
    (2) The ferrule must have no back and forth movement along the tube.
    (3) There must be uniform ridge of metal raised above the tube surface at the end of ferrule.

44. Yes, the bend is acceptable. It could be as flat as 9/16'' and still be acceptable.

45. It has either a smooth or ribbed rubber outer cover.

46. It has a rough cotton braid on its outside.

47. It allows the technician to determine whether or not the hose is twisted.

48. High-pressure hose with a green outer covering is for use with Skydrol fluid only.

49. It must not be straightened or allowed to bend in any way it was not bent when it was installed.

50. Very little. Only a small amount on the second thread from the end.

51. AN flare tube fittings have a small shoulder between the flare cone and the first thread while AC fittings have their threads start at the flare cone.

52. After the first opposition is felt, the fitting should be tightened only 1/6, or, at the most 1/3 turn more.

53. A gasket seals between two stationary surfaces while a packing seals between surfaces where there is relative motion.

54. On the side with the lip.

55. Tightness of the adjusting nut.

56. By using backup rings on the side of the seal away from the pressure.

57. The outside.

58. It would be marked with a green dash.

59. By making sure it has the part number required by the manufacturer, and buying the seal from a reputable dealer.

60. It is used for operations where the same force is required for action in each direction.

61. A rack and pinion type actuator is used.

62. (1) It is instantly reversable.

    (2) There is no fire hazard when it is stalled.

63. A vane type hydraulic motor with vanes on both sides of the shaft driven at the same time.

64. Mechanical energy is turned into heat energy as fluid is forced through a small orifice under pressure.

65. It prevents rebound as the strut attempts to extend.

66. Adding or removing shim washers between the knuckle of the torque links.

67. Adding or removing tapered shims between the axle and the spring steel gear.

68. Compressed air or nitrogen.

69. It either directs main system or emergency system fluid into the actuator.

70. An energizing brake uses the weight of the airplane to aid in the application of the brakes while a nonenergizing brake does not.

71. Nonenergizing.

72. When the brake is applied, the adjuster pin is pulled through its grip. The more the lining wears, the more the pin will pull. When the brake is released, the lining can move back only as far as the adjuster pin will allow.

73. By the amount the automatic adjuster pin sticks out the back of the brake.

74. To prevent the brake dragging from thermal expansion of the fluid.

75. A boosted brake uses system pressure to apply a force on the master cylinder piston while a power brake actually uses system pressure at the wheel cylinder.

76. A pressure regulator.

77. Any time a wheel decelerates to the point that a skid is imminent, the valve releases the pressure to the wheel and allows it to spin back up. It then reapplies the pressure.

78. It decreases the pressure and increases the volume of fluid supplied to the brake when it is applied.

79. A hand-operated control valve meters compressed air to the brake.

80. There is an O-ring between the wheel halves.

81. In the event of excessive overheating from brake application, the fuse material will melt, allowing the tire to deflate rather than blow out.

82. The recommended static load and inflation pressure.

83. The liner of a tubeless tire is rough and will chafe the inner tube.

84. The tire has been operated with too high an inflation pressure.

85. The tire was retreaded the second time in June of 1975. The name of the company is the agency doing the retreading.

86. Some form of dye penetrant inspection.

87. Its light point.

88. This removes any wrinkles and stretches and allows the tube to be fully relaxed when it is finally inflated.

89. These holes vent the plies and allow air trapped between the plies to escape.

90. Electric motors, fluorescent lights, and other electrical devices generate ozone which will cause rubber to deteriorate.

91. Vertically.

92. (1) Isolate the problem.

   (2) Eliminate everything that is good in the problem area. That which you cannot eliminate is where the trouble is.

93. Usually higher.

94. A hissing noise may be heard, and the component may get warm from the bypassing fluid.

95. Pressure bleeding forces all the air in the system to the top as it will naturally want to go, rather than forcing it down as is done with gravity bleeding.

AIRCRAFT HYDRAULICS
Final Examination

STUDENT _____

GRADE _____

Place a circle around the letter for the answer which is most nearly correct.

1. Which statement is true regarding the distribution of pressure in a hydraulic system, according to Pascal's law?

   A. The larger cylinders get the most pressure.
   B. The smaller cylinders get the most pressure.
   C. All cylinders get the same pressure, regardless of their size.
   D. The smaller cylinder will get the most pressure, but the force it exerts will be least.

2. Full pneumatic systems such as found on the Fairchild F-27 uses what for fluid?

   A. Compressed nitrogen carried in high-pressure steel bottles.
   B. Compressed air from engine-driven compressors.
   C. Skydrol fluid.
   D. Compressed air from the maintenance shop air supply.

3. What is the purpose of the spring-loaded valve in many hydraulic filters?

   A. It is a check valve that allows fluid to flow through the filter in only one direction.
   B. Only a certain portion of the system fluid passes through the filter for normal operation. The rest passes through the valve.
   C. It provides enough back pressure for efficient filtering.
   D. It allows fluid to bypass the filter in the event the filter becomes clogged.

4. Which statement is true regarding hydraulic pumps?

   A. A spur gear pump is a constant displacement pump.
   B. An engine-driven constant displacement pump does not require any type of pump control valve or unloading valve.
   C. High-pressure fluid from the discharge side of a gear pump is directed against the side flanges of the gear bushings to prevent air being drawn into the pump from a damaged seal.
   D. A gerotor pump is a variable displacement pump.

5. Open center hydraulic systems:

   A. Have their selector valves arranged in parallel with each other to allow for simultaneous actuation of several systems.
   B. Have their selector valves in series with each other so simultaneous operation cannot be done.
   C. Unload the pump by allowing fluid to return to the reservoir through the open center of the selector valves when a system is not being actuated.
   D. Require an accumulator to maintain pressure when no system is being actuated.

6. What is the purpose of a priority valve?

   A. It determines the priority of component actuation when there is a limited amount of hydraulic fluid in the reservoir.
   B. It automatically selects which system, the main or the emergency, will be used for normal operation.
   C. It is mechanically opened to sequence the actuation of such components as wheel well doors and the landing gear.
   D. It is pressure-actuated to control the sequence of operation of such components as wheel well doors and the landing gear.

7. What is true about automatic hydraulic system pressure regulators?

   A. The pressure on the system is maintained by controlling the size of the orifice in the return line to the reservoir.
   B. The load on the pump is held constant by the regulator whether a system is being actuated or not.
   C. An accumulator is not required with an automatic regulator.
   D. While the accumulator is maintaining system pressure within the operating range of the regulator, the regulator allows the pump to circulate the fluid in the system with little opposition.

8. How should an oleo shock strut be serviced?

   A. Completely collapse the strut and fill it with hydraulic fluid, exercise it to remove any entrapped air, and inflate it with compressed air to an extension specified by the manufacturer.
   B. Extend the strut, fill it with hydraulic fluid, install the air valve, and inflate it to the pressure specified by the manufacturer.
   C. Remove the spring, check its free-standing height, and if it is within tolerance, reinstall it.
   D. Take the weight off of the landing gear and check it to see if the strut piston extends. If it does, tighten the packing nut on the chevron seal.

9. When O-rings are used in a high-pressure component:

   A. A backup ring must be installed on the same side the pressure is applied.
   B. The O-ring must completely fill the groove in the actuator piston.
   C. A backup ring must be installed on the side of the O-ring away from the pressure.
   D. Two O-rings in the same groove will seal better than a single ring.

10. Which statement is true regarding automatic adjusters on a single-disc aircraft brake:

    A. Lining wear is indicated by the length of adjuster pin sticking out of the back of the brake.
    B. The adjuster pin restricts the amount the piston can move in during brake application.
    C. Automatic adjusters trap a small amount of fluid in the wheel cylinder each time the brake is applied.
    D. Automatic adjusters regulate the amount of hydraulic pressure supplied to the cylinder by the master cylinder.

11. What is true about the installation of rigid tubes in a hydraulic system?

    A. Rigid tubing should never be installed between two fittings without at least one bend.
    B. The "B" nut on a flared tube fitting may be used to pull the flare up to the fitting if the distance is no more than one fourth of the tube diameter.
    C. A 3/4 inch diameter tube may be flattened in the bend to 1/2 and still be acceptable.
    D. Hydraulic lines made of aluminum alloy tubing should be installed above electric wire bundles.

12. Wheel alignment is accomplished in an airplane with oleo struts by:

    A. Elongating the bolt holes in the oleo attachment and rotating the strut.
    B. Shimming between the axle and the oleo strut.
    C. The axle is on a cam, and it can be rotated for alignment.
    D. Shimming between the arms of the torque links.

13. Anti-skid brake systems prevent skids by:

    A. Restricting the amount the brake linings can move out against the disc.
    B. Restricting the amount of hydraulic pressure applied to the brake.
    C. Releasing the brake each time it slows down too fast, and reapplying it.
    D. Adjusting the opposition to pedal movement the pilot feels to prevent his applying the brakes too hard.

14. What is the proper way to tighten a flareless tube fitting?

    A. Screw it down by hand until you feel opposition; then turn it down 1-1/2 turns more.
    B. Torque it to the value in foot-pounds recommended in AC 43.13-1A.
    C. Screw it down by hand until you feel opposition; then turn it down 1/2 turn more.
    D. Screw it down by hand until you feel opposition; then turn it down 1/6 to 1/3 turn more.

15. What is the proper way to apply thread lubricant to a tapered pipe fitting?

    A. Apply the lubricant to the bottom male thread only.
    B. Apply the lubricant to the first thread in in the female side and screw the fitting in.
    C. Apply the lubricant sparingly to the second and third male threads and screw the fitting in.
    D. Thread lubricant should never be used with tapered threads, as the taper does the sealing.

16. Why are de-boosters used with power brakes?

    A. They lower the pressure and increase the volume of fluid actually delivered to the brake.
    B. They lower both the pressure and amount of fluid supplied to the brake.
    C. They lower the pressure delivered by the power brake control valve.
    D. They lower the pressure before it goes to the power brake control valve to prevent the brakes grabbing.

17. Air is prevented from leaking from a wheel using a tubeless tire by:

    A. Sealing the wheel halves with a liquid sealant similar to that used for sealing integral fuel tanks.
    B. Using only one piece drop center wheels.
    C. Sealing the wheel halves with an O-ring.
    D. Tubeless tires are not approved for use on airplanes.

18. What is the basic principle of logical troubleshooting?

    A. Isolate the trouble and fix it.
    B. Use a ouija board in the dark of the moon.
    C. Isolate the problem area, then eliminate everything about that area that is working correctly.
    D. Analyze every part of the system that is not working right, beginning with the most complex part.

19. What would likely indicate a system pressure regulator stuck in the open position?

    A. System pressure would drop to zero.
    B. There would be no return of fluid to the reservoir.
    C. The hydraulic system would chatter or hammer.
    D. The pump would tend to overheat.

20. Which statement is **not** true about MIL-H-5606 hydraulic fluid?

    A. This fluid has a mineral base and is colored red.
    B. This fluid is compatible with Skydrol.
    C. Systems using this fluid can be flushed with varsol or naphtha.
    D. Seals used with this fluid are marked with a blue mark.

AVIATION MAINTENANCE PUBLISHERS
P.O. Box 36
Riverton, Wyoming 82501
Toll Free (800)443-9250

FREE   FREE   FREE

This exciting new technical book catalog is *free* for the asking – send for yours today!

The new catalog contains:
* *Numerous technical books for A&P Mechanics and Avionics Technicians*
* *Specialized pilot training books*
* *Many government publications and more!*

Now you can obtain a copy of the popular "Aviation Technical Book Catalog" which contains the complete series of books available from the Aviation Maintenance Publishers. Simply complete this catalog request form and drop it into the mail. No postage necessary.   **So don't delay – send for your FREE AVIATION TECHNICAL BOOK CATALOG TODAY!**

☐ Please send me a **free** copy of your latest technical book catalog.

☐ I would also like to receive a free copy of your Audio-Visual catalog of popular aviation training programs.

Ship to:

Name: _____ Tel. No. _____

Address: _____

City: _____ State: _____ Zip: _____

NO POSTAGE
NECESSARY
IF MAILED
IN THE
UNITED STATES

BUSINESS REPLY MAIL

FIRST CLASS   Permit #8   Riverton, WY 82501

Postage will be paid by:

AVIATION MAINTENANCE PUBLISHERS
P.O. Box 36
1000 College View Drive
Riverton, Wyoming 82501

CATALOG ORDER DEPARTMENT

AIRCRAFT HYDRAULICS
Answers To Final
Examination

1. C
2. B
3. D
4. A
5. C
6. D
7. D
8. A
9. C
10. A
11. A
12. D
13. C
14. D
15. C
16. A
17. C
18. C
19. A
20. B

Notes